Purification and Characterization of Secondary Metabolites

science &
technology books

ELSEVIER

Instructor's Guide Web Site:

https://textbooks.elsevier.com/web/product_details.aspx?isbn=9780128139424

Purification and Characterization of Secondary Metabolites
Thomas E. Crowley

Resources available:

- The laboratory manual that this guide accompanies, Purification and Characterization of Secondary Metabolites: A Laboratory Manual for Analytical and Structural Biochemistry, will be referred to here as "the students' laboratory manual". This guide for the instructor covers only the experimental chapters of the students' laboratory manual, i.e. chapters 9, 10, 11 and 12. The title for each chapter, and for each section within each chapter, is identical to the title of the corresponding chapter or section in the students' laboratory manual.

- Advice for preparing materials for the students to use is presented. For many of the exercises, a detailed record of how I performed the experiments and the data that was obtained is also included.

ELSEVIER

ACADEMIC
PRESS

Purification and Characterization of Secondary Metabolites

A Laboratory Manual for Analytical and Structural Biochemistry

Thomas E. Crowley
Department of Mathematics & Natural Sciences,
National University, La Jolla, CA, United States

ACADEMIC PRESS
An imprint of Elsevier

Academic Press is an imprint of Elsevier
125 London Wall, London EC2Y 5AS, United Kingdom
525 B Street, Suite 1650, San Diego, CA 92101, United States
50 Hampshire Street, 5th Floor, Cambridge, MA 02139, United States
The Boulevard, Langford Lane, Kidlington, Oxford OX5 1GB, United Kingdom

Notices
Knowledge and best practice in this field are constantly changing. As new research and experience broaden
our understanding, changes in research methods, professional practices, or medical treatment may become
necessary.

Practitioners and researchers must always rely on their own experience and knowledge in evaluating and
using any information, methods, compounds, or experiments described herein. In using such information or
methods they should be mindful of their own safety and the safety of others, including parties for whom they
have a professional responsibility.

To the fullest extent of the law, neither the Publisher nor the authors, contributors, or editors, assume any
liability for any injury and/or damage to persons or property as a matter of products liability, negligence or
otherwise, or from any use or operation of any methods, products, instructions, or ideas contained in the
material herein.

British Library Cataloguing-in-Publication Data
A catalogue record for this book is available from the British Library

Library of Congress Cataloging-in-Publication Data
A catalog record for this book is available from the Library of Congress

ISBN: 978-0-12-813942-4

For Information on all Academic Press publications
visit our website at https://www.elsevier.com/books-and-journals

Publisher: Andre Gerhard Wolff
Acquisition Editor: Peter Linsley
Editorial Project Manager: Sam Young
Production Project Manager: Sreejith Viswanathan
Cover Designer: Matthew Limbert

Typeset by MPS Limited, Chennai, India

Working together
to grow libraries in
developing countries

www.elsevier.com • www.bookaid.org

Contents

Preface

This book is written for students enrolled in an undergraduate- or graduate-level biochemical laboratory course. The students participating in such a course should have completed lecture and laboratory courses in both general and organic chemistry. In addition to being a laboratory manual, this book is also a source of knowledge pertaining to the theoretical basis of microbiological and biochemical techniques. Discussion of the current knowledge of the structure and biological function of the metabolites to be studied in these exercises is also included. This aspect of the book is supported by an extensive bibliography.

The early chapters provide guidance in the methods for the growth of microbial cultures and purification of secreted secondary metabolites. The theoretical basis is presented for methods to quantify the concentration, assay the intermolecular interactions, and elucidate the structures of metabolites. Later chapters consist of exercises in which secondary metabolites from three species of bacteria are purified and their structures characterized. These exercises include detailed protocols that help students who are just beginning to learn biochemical skills to acquire quality data. The bacteria from which the metabolites are derived represent a wide range of microbial characteristics. All of the procedures in the exercises have been tested and been found to be reliable. The final chapter provides suggestions for the design of original experiments in which a previously uncharacterized fluorescent metabolite is purified from a bacterial culture. Guidance in designing experiments to characterize the structure of this fluorescent metabolite is also included in this chapter.

A Instructor's Guide website for this book is available at: https://text-books.elsevier.com/web/product_details.aspx?isbn=9780128139424.

Acknowledgments

In the Department of Mathematics and Natural Sciences at National University

Professor Veronica Ardi-Pastores and Juliann Downing initiated the acquisition of the preparative centrifuge that is necessary for the initial step in the purification of a metabolite from a bacterial culture. Professor Jacqueline Ruiz Harewood provided advice and assistance pertaining to extraction of the bacterial cultures with organic solvents. Juliann was also extremely helpful with a variety of technical issues as I tested the various protocols included in this book. Her assistance with the infrared spectrometric, ultraviolet–visible spectrophotometric and high-performance liquid chromatographic devices allowed me to collect high-quality data. She also provided help with the use of the laboratory equipment and supplies that are necessary for the growth of bacterial cultures.

At Other Institutions

Professor Arnold Rheingold of the Department of Chemistry and Biochemistry, at the University of California at San Diego, provided instruction in X-ray crystallography. Arnold and Curtis Moore assisted me in acquisition and interpretation of crystallographic data at the Crystallography Facility at UCSD. Anthony Mrse at the Nuclear Magnetic Resonance Facility in the Department of Chemistry and Biochemistry at UCSD provided instruction and technical assistance. Bill Webb at the Scripps Center for Metabolomics at the Scripps Research Institute in La Jolla was helpful with technical issues that allowed me to collect good data from assays with mass spectrometry.

Family

Most importantly, my wife, Stella Crowley, was a constant source of support while this work was being completed. Her patience and understanding during the past two years allowed me to properly focus my energy while I was writing this book.

Chapter 1

Working safely and efficiently in the biochemical laboratory

1.1 WORKING SAFELY WITH MICROORGANISMS

Fundamental restrictions and sources of information pertinent to biosafety

Eating and drinking are not allowed in a laboratory in which experiments with microorganisms are performed.

The Centers for Disease Control and Prevention

https://www.cdc.gov/biosafety/

This agency defines four *biosafety levels* for work with microorganisms in the laboratory. The following are excerpts of the agency's description of the two levels that are relevant to the exercises included in this book.

Biosafety Level 1 (BSL-1).

Biosafety Level 1 practices, safety equipment, and facility design and construction are appropriate for undergraduate and secondary educational training and teaching laboratories, and for other laboratories in which work is done with defined and characterized strains of viable microorganisms not known to consistently cause disease in healthy adult humans.

BSL-1 represents a basic level of containment that relies on standard microbiological practices with no special primary or secondary barriers recommended, other than a sink for hand washing.

Biosafety Level 2 (BSL-2).

Biosafety Level 2 practices, equipment, and facility design and construction are applicable to clinical, diagnostic, teaching, and other laboratories in which work is done with the broad spectrum of indigenous moderate-risk agents that are present in the

Purification and Characterization of Secondary Metabolites.
DOI: https://doi.org/10.1016/B978-0-12-813942-4.00001-2
© 2020 Elsevier Inc. All rights reserved.

community and associated with human disease of varying severity. With good microbiological techniques, these agents can be used safely in activities conducted on the open bench, provided the potential for producing splashes or aerosols is low.

Personal protective equipment should be used as appropriate, such as splash shields, face protection, gowns, and gloves. Secondary barriers, such as hand washing sinks and waste decontamination facilities, must be available to reduce potential environmental contamination.

The American Type Culture Collection

https://www.atcc.org/Support/How_to_Order/Biosafety_Level.aspx.

For each bacterial strain in the collection, a *product sheet* is posted on this website. This document specifies the biosafety level that is appropriate for handling the strain.

The bacterial strains to be used in the exercises in this book

The American Type Culture Collection states that the MJ1 or MJ11 strain of *Vibrio fischeri*, and the glycinea pathovar of *Pseudomonas syringae*, may be safely handled with the BSL-1 guidelines. It has also been noted in the scientific literature that the commercially available strains of *Enterobacter aerogenes* may be safely handled with the BSL-1 guidelines (Johnson & Case, 2018).

Your instructor may prefer to use the BSL-2 guidelines as an extra precaution. The most significant difference is that in BSL-2 at the end of the experiment all bacterial cultures, and material contaminated with bacteria, must be considered a biohazard. These items must be decontaminated with high temperature or chemicals, or placed in a hard plastic container designated for sharp biohazardous waste.

1.2 WORKING SAFELY WITH CHEMICALS AND LABORATORY EQUIPMENT

Fundamental restrictions and administrative issues

Eating and drinking are not allowed in a biochemical laboratory.

Department of Environmental Health and Safety

Most colleges and universities have a department with this title or a similar title. If there is not a department with this title, there should be one or more safety officers. The staff of this department create and enforce policies. They assure that safety equipment such as chemical fume hoods, safety showers, and stations for the washing of eyes are functioning properly. They also handle disposal of hazardous waste and response to emergencies.

Personal protective equipment

Protection for your eyes

Whenever chemicals are being handled in the laboratory, everyone in the laboratory should be wearing protection for their eyes. For most of the procedures described in this book, plastic *safety glasses* are sufficient. During liquid–liquid or solid-phase extraction, there is a possibility of splashing of large volumes of hazardous liquids. For such procedures, *safety goggles* should be worn.

Protection for your arms and body

Whenever chemicals are being handled in the laboratory, everyone in the laboratory should be wearing a laboratory coat. The sleeves of the coat should extend to the wrists.

Protection for your feet

Everyone in the laboratory must be wearing closed-top shoes that do not have high heels.

Protection for your hands

For some procedures your instructor or the teaching assistant may recommend that you wear disposable gloves. Brief summaries of the most significant hazards of the chemicals used in the exercises in this book are provided in Table 1.1. If you are allergic to latex, wear nitrile or other hypoallergenic gloves.

Hazards from equipment

The most hazardous type of equipment used in the purification of secondary metabolites is the preparative centrifuge. A typical preparative centrifuge is designed to process as much as 4.0 L

TABLE 1.1 The hazards of some of the chemicals used in the exercises in this book.

Chemical[a]	Procedure	Hazard[b]
Formic acid	Liquid chromatography	Flammable, burns skin, damages eyes, irritates respiratory tract
Acetic acid	Liquid–liquid extraction	Flammable, burns skin, damages eyes, irritates respiratory tract
Hydrochloric acid	Liquid–liquid extraction	Severe damage to skin and eyes, irritates respiratory tract
Trifluoro-acetic acid	Liquid chromatography	Burns skin, damages eyes, irritates respiratory tract
Ammonium hydroxide (aqueous)	Purification of siderophore	Burns skin, damages eyes, toxic if swallowed
Methanol	Solid-phase extraction	Highly flammable, damages eyes, irritates respiratory tract, toxic if inhaled
Acetonitrile	Liquid chromatography	Highly flammable, irritates eyes and skin
Ethyl acetate	Liquid–liquid extraction	Highly flammable, irritates eyes, causes dizziness, toxic if inhaled
Dimethyl sulfoxide	Nuclear magnetic resonance spectroscopy	Combustible and flammable, rapidly adsorbed through skin

[a]All of these chemicals are used in the liquid phase.
[b]From the safety data sheets provided by MilliporeSigma.

of liquid samples with a maximum revolution per minute (rpm) of 4200. Because of the large size of the rotor and buckets that revolve around the spindle, the following precautions must be taken.

• Samples should be in screw-cap tubes or bottles that are designed for the amount of centrifugal force that will be applied.
• A two-pan balance should be used to adjust samples that will be in buckets on opposite sides of the rotor to an equal mass before loading. The tubes or bottles will be open while balancing (to allow for addition of more solution) but the caps should be on the trays of the balance. If adaptors are to be used to make the tubes or bottles fit snugly in the buckets, they should also be on the trays.

- In most cases, the rotor is only secured to the spindle of the centrifuge if the cap of the rotor has been screwed on tightly.
- After starting the centrifugation, stay near the centrifuge until it reaches the desired number of revolutions per minute. If severe vibration of the machine occurs, stop the centrifugation immediately!

Safety data sheet for a chemical

This document provides warnings regarding the combustibility, flammability, corrosiveness, and toxicity of a chemical.

The public website that is maintained by MilliporeSigma provides the safety data sheet (SDS) for many chemicals.

https://www.sigmaaldrich.com/safety-center.html

The *Safety Emporium* is a public website that is maintained by *Interactive Learning Paradigms Incorporated* (ILPI). This website provides links to many sources of SDS.

http://www.ilpi.com/msds/index.html

Brief summaries of the hazards from chemicals that will be used in the exercises in this book are provided here (Table 1.1).

Terminology that pertains to toxic chemicals

Carcinogen

A chemical that has the potential to induce genetic mutations that may induce cancer.

Fetotoxin

A chemical that is toxic to a fetus.

Teratogen

A chemical that may induce genetic mutations in a fetus that may result in developmental abnormalities.

Minimizing exposure to hazardous chemicals

Volatile chemicals

To ensure that workers in the laboratory do not inhale toxic vapors, containers holding volatile chemicals should only be opened inside of a chemical fume hood. Disposal of volatile chemical waste should be in a container that is stored within the hood and is designated for this purpose.

Corrosive, toxic, or carcinogenic chemicals

Wear disposable gloves when handling these chemicals. If any are spilled on your skin, rinse immediately with water.

Responding to an emergency in the laboratory

Note: These instructions are for the instructor and teaching assistant, as well as for the student.

Calling for help

If it appears necessary, a call for professional emergency medical help should be made. Before the first day of class, learn the campus policy covering procedures for such calls. On some campuses, the policy may be to call 9-1-1 directly from the laboratory. At other schools you may be required to call the department of environmental health and safety or campus health services so that they can call 9-1-1 and direct the responders to the desired location.

Contact of skin or eyes with hazardous chemicals

The laboratory should have emergency showers and eyewash stations at the sinks. The instructor, teaching assistants, and students should be aware of their location and how to use them. If necessary, call the National Capital Poison Center at 800-222-1222. The URL of the center's website is https://www.poison.org.

Injury

The laboratory should have a first-aid kit. Everyone working in the laboratory should know where it is located. If an instructor, teaching assistant, or student believes he or she has the necessary understanding to treat a minor injury, he or she should assist in the treatment.

Sudden illness such as fainting or nausea

Move the affected individual out of the laboratory and away from chemical vapors or other hazards. If the recovery is not rapid and the individual's condition seems serious, call for professional medical help as described earlier.

Spill of a chemical

There should be a spill kit in the laboratory. The instructor or the teaching assistant should teach you how to use it. It should contain the following materials:

- sodium carbonate or sodium bicarbonate for spills of acid;
- citric acid or sodium bisulfate for spills of alkali; and
- inert adsorbents such as vermiculite, clay, or sand for hazardous liquids.

Fire

The instructor, teaching assistants, and students should know the location of the fire extinguishers and be prepared to use them. Learn the evacuation plan on the first day of class.

Earthquake

If you are in a region that is prone to earthquakes, learn the recommended response on the first day of class.

1.3   PREPARING FOR EXPERIMENTS, PERFORMING CALCULATIONS, AND RECORDING DATA

Your laboratory notebook

Before the first session of laboratory work you should purchase a bound notebook with numbered pages. For the duration of the course, no pages should be removed from this notebook. All writing in the notebook should be with ink, not pencil. Mistakes and unnecessary items should be crossed out rather than erased.

Before each session, a brief summary of the experimental plan should be written in the notebook. If appropriate, the summary should be followed by mathematical calculations that are necessary to perform the experiment. Templates of tables for the recording of data may also be appropriate for some experiments. Because this part of the experimental record is performed before the session begins, it should be neat and well organized.

As the experiment is being performed, more calculations may be necessary. These calculations should also be written in the notebook. In addition, observations and data should be written in the notebook. The written record of your work during the session in the laboratory is not expected to be extremely neat and well organized. Because the work during the session must be performed quickly, it is impractical to make this part of the notebook highly refined. Most instructors require students to submit *laboratory reports*. Reports are usually written after the session in the laboratory has ended, thus they are expected to be more refined than the written record in the notebook.

Personal electronic devices

The only personal electronic device that a student should use in a laboratory is a pocket calculator. Portable computers, phones, and radios should not be powered on during the session.

BIBLIOGRAPHY

Johnson, T. R., & Case, C. L. (2018). *Laboratory experiments in microbiology* (12th ed.). Hoboken: Pearson. Available from https://lccn.loc.gov/2017039734.

Chapter 2

The structure and function of secondary metabolites that are secreted by bacteria

2.1 THE SIGNIFICANCE OF SECONDARY METABOLITES

Biochemists and microbiologists refer to many of the biological molecules that have a molar mass of $<2000.0\,\mathrm{g\,mol}^{-1}$ as *metabolites*. A *primary metabolite* is typically an intermediate in a fundamental metabolic pathway, such as glycolysis or the citric acid cycle. The structures of the primary metabolites do not vary much between biological species, or between different types of cells in a single species. The function of most of the primary metabolites is inside the cell.

In contrast, most *secondary metabolites* are more specialized in structure and function. Each of these substances is expressed by only one, or a few, species. Many of them are secreted from the cells in which they are synthesized. The subsequent sections in this chapter provide introductions to three secondary metabolites, each of which is secreted by a distinct species of bacteria. These organic molecules have a wide range of biological functions.

2.2 A SIGNALING MOLECULE SECRETED BY *VIBRIO FISCHERI* FUNCTIONS IN QUORUM SENSING

Note: Some scientists are now classifying the bacterial species *fischeri* in the genus *Aliivibrio*. For the reasons explained in *Theoretical Background* in the later chapter in which exercises with this species are presented, the genus will be specified as *Vibrio* in this book.

Purification and Characterization of Secondary Metabolites.
DOI: https://doi.org/10.1016/B978-0-12-813942-4.00002-4
© 2020 Elsevier Inc. All rights reserved.

Bioluminescence and quorum sensing in bacteria

Bioluminescence and quorum sensing are two of the most fascinating properties of bacteria. The natural environment of most of the luminescent bacterial species is the ocean (Dunlap & Kita-Tsukamoto, 2006). Apparently they are better adapted to the high ionic strength of seawater than the low ionic strength of freshwater in lakes and streams. These microbes may be found floating freely in a *planktonic* mode of growth, or in an aggregate that is embedded within a tissue in a fish (Tortora, Funke, & Case, 2019; Winfrey, Rott, & Wortman, 1997). The growth within the fish is a *symbiosis* because the luminescence promotes the survival of the host. Symbiotic growth generates a much higher density of bacterial cells than does planktonic growth.

In many types of symbiotic growth, bacteria communicate by quorum sensing (Bassler & Miller, 2006). A secondary metabolite that functions as a signaling molecule is secreted by some of the bacteria and then internalized by other bacteria of the same species. The bacteria that have received this signal respond by synthesizing and secreting the same type of signal. This newly synthesized signal then propagates the induction to more bacteria. The amplification of the induction is most effective in symbiotic growth because of the proximity of the bacteria to one another.

The luminescent bacterium *Vibrio fischeri* grows planktonically in seawater and in symbiosis with several species of fish (Kimbrough & Stabb, 2015; Thompson et al., 2017). The intensity of the luminescence in this bacterial species increases dramatically when the density of cells increases, thus it is usually observed in symbiotic growth. Intercellular communication by means of quorum sensing induces the luminescence.

The lux operon and *Vibrio* autoinducer 1

Expression of the *lux* operon in the genome of *V. fischeri* is the link between quorum sensing and luminescence in this bacterium (Stabb & Visick, 2013; Winfrey et al., 1997). The gene *luxI* encodes an enzyme that synthesizes the signaling molecule *Vibrio* autoinducer 1 (VAI-1). The receptor for this signal is the polypeptide encoded by the gene *luxR*. The symbol for this polypeptide is LuxR. Other genes within this operon encode the enzymes that catalyze the luminescence-generating reactions. After VAI-1 enters a *V. fischeri* bacterium, it binds to

LuxR. The complex of receptor plus ligand then makes contact with a *cis*-regulatory sequence of DNA within the *lux* operon, inducing transcription of the genes for luminescence.

2.3   SIDEROPHORES ALLOW BACTERIA TO ACQUIRE IRON FROM THEIR ENVIRONMENT

Almost all types of cells in all species of organisms require iron for survival. The roles of iron(II) and iron(III) (also known as Fe^{2+} and Fe^{3+}) vary between different types of cells and species. One of the most important functions is as a prosthetic group in the hemoproteins (cytochromes) in the electron transport-oxidative phosphorylation system in cells that perform aerobic or anaerobic respiration.

Because ionic iron is not very abundant in the various environments in which bacteria grow, a mechanism for chelation and importation of iron(III) has evolved in these species. Secondary metabolites known as *siderophores* that are secreted by these bacteria chelate ions of iron(III) and the organometallic complex is then imported into the cell. In such a complex the organic portion is referred to as the *ligand*. Several types of siderophores are discussed in the following paragraphs. Sample data characterizing two of these iron chelators will be presented in later chapters. The other two will be the subject of experimental exercises in later chapters.

The first siderophore to be thoroughly characterized was enterobactin, which is expressed by several species of bacteria including *Escherichia coli* (Pollack & Neilands, 1970; Raymond & Carrano, 1979; Raymond, Dertz, & Kim, 2003). Enterobactin is a cyclic molecule with three catechol groups that generate a hexadentate complex with iron(III) in an octahedral coordination. Because iron is not associated with a siderophore while it is being synthesized and secreted, this form of the metabolite is known as *desferric* enterobactin. After chelation occurs it is referred to as *ferric* enterobactin. This nomenclature is used for some of the other siderophores. As is explained in the next paragraph, for the several types of *ferrioxamine* a distinct system is used.

A different functional group that interacts with the iron in some siderophores is hydroxamic acid. The negatively charged hydroxamate that is formed by ionization of this acid provides the chelating activity. Examples of siderophores containing hydroxamate include ferrioxamine B, ferrioxamine E, and aerobactin. Ferrioxamine B is expressed by *Streptomyces pilosus* (Borgias, Hugi, & Raymond, 1989; Dhungana, White, & Crumbliss, 2001) and multiple species in the genus *Micromonospora* (Simionato et al., 2006). Ferrioxamine E is expressed by *Streptomyces antibioticus* (Van der Helm & Poling, 1976). Aerobactin is expressed by *Klebsiella pneumoniae* (Küpper, Carrano, Kuhn, & Butler, 2006), *Enterobacter cloacae* (Keller, Pedroso, Ritchmann, & Silva, 1998; Van Tiel-Menkveld, Mentjox-Vervuurt, Oudega, & de Graaf, 1982), and a species in the genus *Vibrio* (Haygood, Holt, & Butler, 1993). The nomenclature explained earlier for specifying whether or not enterobactin is bound to iron(III) is also used for aerobactin; however, for each ferrioxamine, a prefix is used. Ferrioxamine B is the ligand plus iron(III), *des*ferrioxamine B is the ligand by itself.

An aqueous solution containing iron(III) appears pale yellow. After being chelated by a ligand such as a siderophore, the color changes to orange, red, brown, or blue (Gibson & Magrath, 1969; Küpper et al., 2006; Pecoraro, Harris, Wong, Carrano, & Raymond, 1983; Young & Gibson, 1979). The new color depends on the structure of the siderophore and the pH of the solution. This phenomenon provides a convenient technique for detecting the presence of a siderophore in a solution (Fig. 2.1).

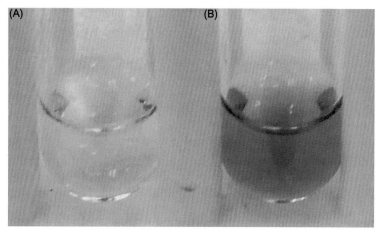

FIGURE 2.1 Enterobactin, a siderophore secreted by *Escherichia coli*, changes the color of iron in an aqueous solution. (A) Crystals of iron(III) chloride were dissolved in an aqueous solution of 5.0 mM HCl, releasing iron(III) at 120.0 mM. The iron(III) makes the solution appear yellow. (B) Addition of enterobactin to 75.0 μM to the acidic solution of $FeCl_3$ (aq) results in formation of an organometallic complex of enterobactin and iron(III). The solution now appears brown. The enterobactin was obtained from a commercial supplier (Sigma-Aldrich_Co_enterobactin, 2017).

2.4   CORONATINE IS SECRETED BY PATHOVARS OF *PSEUDOMONAS SYRINGAE* AND IS TOXIC TO CHLOROPLASTS IN CROP PLANTS

Phytopathogenic bacteria

A *saprophyte* is a bacterium that establishes a symbiosis with a plant. A *phytopathogen* is a bacterium that infects a plant and causes disease. The systematic name for a bacterium is two words, the first representing the genus and the second the species (Johnson & Case, 2018; Tortora et al., 2019). Both words are written in oblique font. The initial letter in the name of the genus is in uppercase, whereas all letters in the name of the species are in lowercase. Genetic variation that occurs within each species results in subdivision into various *strains*. Each strain within a phytopathogenic bacterial species may be referred to as a *pathovar*. The abbreviation "pv," not in oblique font, is sometimes used. Each pathovar within a bacterial species infects a different species of plant (Alfano & Collmer, 1996). Examples of two pathovars within a bacterial species are *Erwinia carotovora* pv tomato and *E. carotovora* pv tobacco.

Chlorosis may be induced by a toxin that is secreted by the bacteria associated with a plant. Chlorosis in a leaf involves shrinkage of the chloroplasts (Fig. 2.2) and degradation of the pigment chlorophyll within these organelles.

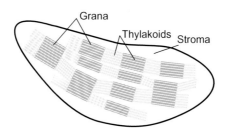

FIGURE 2.2   The chloroplast is the subcellular organelle in the leaf of a plant that is affected by chlorosis. Each *thylakoid* is a membrane-enclosed compartment and the interior of this compartment is known as the lumen. The thylakoids contain the green pigment chlorophyll. The term *grana* refers to a stack of thylakoids. The portion of the chloroplast outside of the thylakoids is known as the *stroma*. *Reprinted with permission from Crowley, T. E., & Kyte, J. (2014). Experiments in the purification and characterization of enzymes: A laboratory manual. Amsterdam; Boston: Elsevier/AP, Academic Press is an imprint of Elsevier. https://lccn.loc.gov/2014395777.*

Coronatine

Pathovars of the phytopathogen *Pseudomonas syringae* secrete a variety of secondary metabolites that are toxic to plant cells. These toxins are part of the mechanism by which the bacteria infect the plants. One of these toxins is coronatine and it contributes to infection in crop plants. Coronatine is secreted by *P. syringae* pathovar glycinea (Palmer & Bender, 1993) and *P. syringae* pathovar tomato (Brooks et al., 2004). Pathovar glycinea infects *Glycine max* (soybean).

The systematic name of tomato is *Solanum lycopersicum*. Plants such as *G. max* and *S. lycopersicum* have a signal transduction pathway that defends them against infection. A key component of this pathway is *salicylic acid*. Plants control the activity of this pathway by synthesizing a hormone, jasmonoyl isoleucine (JA-Ile). This hormone allows the plant to block the salicylic acid pathway when it is not needed, thus preventing a waste of the organism's resources.

Two aspects of the function of coronatine have been well studied. One is the interference with the salicylic acid pathway (Panchal et al., 2016; Uppalapati et al., 2007) and the other is the induction of chlorosis (Palmer & Bender, 1995). The structure of coronatine resembles that of JA-Ile. The predominant theory for the action of coronatine on the salicylic acid pathway is that it mimics the inhibitory action of JA-Ile.

BIBLIOGRAPHY

Alfano, J. R., & Collmer, A. (1996). Bacterial pathogens in plants: Life up against the wall. *Plant Cell*, *8*, 1683–1698. Available from https://www.ncbi.nlm.nih.gov/pubmed/12239358.

Bassler, B. L., & Miller, M. B. (2006). Quorum sensing. In M. Dworkin, S. Falkow, E. Rosenberg, K.-H. Schleifer, & E. Stackebrandt (Eds.), *The prokaryotes: Vol. 2: Ecophysiology and biochemistry* (pp. 336–353). New York, NY: Springer. Available from https://doi.org/10.1007/0-387-30742-7_12.

Borgias, B., Hugi, A. D., & Raymond, K. N. (1989). Isomerization and solution structures of desferrioxamine B complexes of aluminum(3 +) and gallium (3 +). *Inorganic Chemistry*, *28*, 3538–3545. Available from https://doi.org/10.1021/ic00317a029.

Brooks, D. M., Hernández-Guzmán, G., Kloek, A. P., Alarcón-Chaidez, F., Sreedharan, A., Rangaswamy, V., ... Kunkel, B. N. (2004). Identification and characterization of a well-defined series of coronatine biosynthetic mutants of *Pseudomonas syringae* pv. tomato DC3000. *Molecular Plant-Microbe Interaction*, *17*, 162–174. Available from https://www.ncbi.nlm.nih.gov/pubmed/14964530.

Crowley, T. E., & Kyte, J. (2014). *Experiments in the purification and characteri-zation of enzymes: A laboratory manual.* Amsterdam; Boston: Elsevier/AP, Academic Press is an imprint of Elsevier. Available from https://lccn.loc.gov/ 2014395777.

Dhungana, S., White, P. S., & Crumbliss, A. L. (2001). Crystal structure of fer-rioxamine B: A comparative analysis and implications for molecular recogni-tion. *Journal of Biological Inorganic Chemistry, 6,* 810−818. Available from https://www.ncbi.nlm.nih.gov/pubmed/11713688.

Dunlap, P. V., & Kita-Tsukamoto, K. (2006). Luminous bacteria. In M. Dworkin, S. Falkow, E. Rosenberg, K.-H. Schleifer, & E. Stackebrandt (Eds.), *The pro-karyotes: Vol. 2: Ecophysiology and biochemistry* (pp. 863−892). New York, NY: Springer. Available from https://doi.org/10.1007/0-387-30742-7_27.

Gibson, F., & Magrath, D. I. (1969). The isolation and characterization of a hydroxamic acid (aerobactin) formed by Aerobacter aerogenes 62-I. *Biochim Biophys Acta, 192,* 175−184. Available from https://www.ncbi.nlm. nih.gov/pubmed/4313071.

Haygood, M. G., Holt, P. D., & Butler, A. (1993). Aerobactin production by a planktonic marine *Vibrio* sp. *Limnology and Oceanography, 38,* 1091−1097. Available from https://doi.org/10.4319/lo.1993.38.5.1091.

Johnson, T. R., & Case, C. L. (2018). *Laboratory experiments in microbiology* (12th ed.). Hoboken: Pearson. Available from https://lccn.loc.gov/ 2017039734.

Keller, R., Pedroso, M. Z., Ritchmann, R., & Silva, R. M. (1998). Occurrence of virulence-associated properties in *Enterobacter cloacae. Infection and Immunity, 66,* 645−649. Available from http://iai.asm.org/content/66/2/ 645.abstract.

Kimbrough, J. H., & Stabb, E. V. (2015). Antisocial luxO mutants provide a stationary-phase survival advantage in *Vibrio fischeri* ES114. *Journal of Bacteriology, 198,* 673−687. Available from https://www.ncbi.nlm.nih.gov/ pubmed/26644435.

Küpper, F. C., Carrano, C. J., Kuhn, J. U., & Butler, A. (2006). Photoreactivity of iron(III)-aerobactin: photoproduct structure and iron(III) coordination. *Inorganic Chemistry, 45,* 6028−6033. Available from https://www.ncbi.nlm. nih.gov/pubmed/16842010.

Palmer, D., & Bender, C. (1993). Effects of environmental and nutritional factors on production of the polyketide phytotoxin coronatine by *Pseudomonas syr-ingae* pv. Glycinea. *Applied and Environmental Microbiology, 59,* 1619−1626. Available from http://www.ncbi.nlm.nih.gov/entrez/query.fcgi? cmd = Retrieve&db = PubMed&dopt = Citation&list_uids = 16348941.

Palmer, D. A., & Bender, C. L. (1995). Ultrastructure of tomato leaf tissue trea-ted with the pseudomonad phytotoxin coronatine and comparison with methyl jasmonate. *Molecular Plant-Microbe Interactions, 8,* 683−692. Available from http://www.apsnet.org/publications/mpmi/BackIssues/ Documents/1995Abstracts/Microbe08-683.htm.

Panchal, S., Roy, D., Chitrakar, R., Price, L., Breitbach, Z. S., Armstrong, D. W., & Melotto, M. (2016). Coronatine facilitates *Pseudomonas syringae* infection of arabidopsis leaves at night. *Frontiers in Plant Science, 7,* 880. Available from https://www.ncbi.nlm.nih.gov/pubmed/27446113.

Pecoraro, V. L., Harris, W. R., Wong, G. B., Carrano, C. J., & Raymond, K. N. (1983). Coordination chemistry of microbial iron transport compounds. 23. Fourier transform infrared spectroscopy of ferric catechoylamide analogues of enterobactin. *Journal of the American Chemical Society*, *105*, 4623–4633. Available from https://doi.org/10.1021/ja00352a018.

Pollack, J. R., & Neilands, J. B. (1970). Enterobactin, an iron transport compound from *Salmonella typhimurium*. *Biochemical and Biophysical Research Communications*, *38*, 989–992. Available from http://www.sciencedirect.com/science/article/pii/0006291X70908193.

Raymond, K. N., & Carrano, C. J. (1979). Coordination chemistry and microbial iron transport. *Accounts of Chemical Research*, *12*, 183–190. Available from https://doi.org/10.1021/ar50137a004.

Raymond, K. N., Dertz, E. A., & Kim, S. S. (2003). Enterobactin: An archetype for microbial iron transport. *Proceedings of the National Academy of Sciences*, *100*, 3584–3588. Available from http://www.pnas.org/content/100/7/3584.abstract.

Sigma-Aldrich_Co_enterobactin (2017) Enterobactin from Escherichia coli, item E3910. St Louis, MO. http://www.sigmaaldrich.com/catalog/product/sigma/e3910?lang = en&region = US.

Simionato, A., de Souza, G., Rodrigues-Filho, E., Glick, J., Vouros, P., & Carrilho, E. (2006). Tandem mass spectrometry of coprogen and deferoxamine hydroxamic siderophores. *Rapid Communications in Mass Spectrometry*, *20*, 193–199. Available from http://www.ncbi.nlm.nih.gov/entrez/query.fcgi?cmd = Retrieve&db = PubMed&dopt = Citation&list_uids = 16345131.

Stabb, E. V., & Visick, K. L. (2013). Vibrio fisheri: Squid symbiosis. In E. Rosenberg, E. F. DeLong, S. Lory, E. Stackebrandt, & F. Thompson (Eds.), *The prokaryotes: Prokaryotic biology and symbiotic associations* (pp. 497–532). (Berlin, Heidelberg: Springer. Available from https://doi.org/10.1007/978-3-642-30194-0_22.

Thompson, L. R., Nikolakakis, K., Pan, S., Reed, J., Knight, R., & Ruby, E. G. (2017). Transcriptional characterization of *Vibrio fischeri* during colonization of juvenile *Euprymna scolopes*. *Environmental Microbiology*, *19*, 1845–1856. Available from https://www.ncbi.nlm.nih.gov/pubmed/28152560.

Tortora, G. J., Funke, B. R., & Case, C. L. (2019). *Microbiology: An introduction* (13th ed). Boston: Pearson. Available from https://lccn.loc.gov/2017044147.

Uppalapati, S. R., Ishiga, Y., Wangdi, T., Kunkel, B. N., Anand, A., Mysore, K. S., & Bender, C. L. (2007). The phytotoxin coronatine contributes to pathogen fitness and is required for suppression of salicylic acid accumulation in tomato inoculated with *Pseudomonas syringae* pv. tomato DC3000. *Molecular Plant-Microbe Interactions*, *20*, 955–965. Available from https://www.ncbi.nlm.nih.gov/pubmed/17722699.

Van der Helm, D., & Poling, M. (1976). The crystal structure of ferrioxamine E. *Journal of the American Chemical Society*, *98*, 82–86. Available from https://doi.org/10.1021/ja00417a014.

Van Tiel-Menkveld, G. J., Mentjox-Vervuurt, J. M., Oudega, B., & de Graaf, F. K. (1982). Siderophore production by *Enterobacter cloacae* and a common receptor protein for the uptake of aerobactin and cloacin DF13. *Journal of*

Bacteriology, *150*, 490−497. Available from http://jb.asm.org/content/150/2/490.abstract.

Winfrey, M. R., Rott, M. A., & Wortman, A. T. (1997). *Unraveling DNA: Molecular biology for the laboratory*. Upper Saddle River, NJ: Prentice-Hall. Available from https://lccn.loc.gov/96036610.

Young, I. G., & Gibson, F. (1979). Isolation of enterochelin from *Escherichia coli*. *Methods in Enzymology*, *56*, 394−398. Available from http://www.science-direct.com/science/article/pii/0076687979560376.

Chapter 3

Overview of the methods for purification of metabolites that are secreted by bacteria

3.1 INOCULATION AND INCUBATION OF BACTERIAL CULTURES

Literature pertaining to bacteriology

There are several sources that discuss the characteristics of bacteria that are pertinent to the growth of cultures in a laboratory. There are two types of bacteriology references known as the *Bergey's Manual*. Nine editions of the original, *Bergey's Manual of Determinative Bacteriology*, have been printed. The most recent of these was published in 1994 (Holt, Krieg, Sneath, Staley, & Williams, 1994). More recently, the *Bergey's Manual of Systematics of Archaea and Bacteria* was published. The current version of this source is only available in digital format on the Internet (Whitman, 2015). A reference book for choosing appropriate medium for the culture of bacteria has been published (Atlas, 2010) and two microbial knowledge bases may be accessed on the Internet (List_of_Prokaryotic_Names_with_Standing_in_Nomenclature, 2018; Strain_Info, 2018). A textbook designed to accompany a lecture course providing comprehensive coverage of the characteristics of bacteria and other microbes (Tortora, Funke, & Case, 2019) and a laboratory manual for a course in bacteriology (Johnson & Case, 2018) are also excellent references.

Choice of medium for the culture

A *complex medium* is one that includes an extract of tissue from an animal, plant, or fungus (Atlas, 2010; Johnson & Case, 2018; Tortora et al., 2019). Some examples of the components found in complex media are tryptone, peptone, and an extract

Purification and Characterization of Secondary Metabolites.
DOI: https://doi.org/10.1016/B978-0-12-813942-4.00003-6

of yeast. These media are considered to be complex because the exact chemical composition of such extracts is not known. Although most species of microbes grow well on one of the complex media, for any given species there are probably many substances in the complex medium that are not essential for growth.

In contrast, a *chemically defined medium* is one in which the exact concentration of each component is known. Only components that are either essential for the growth of the bacterium or necessary to modulate some aspect of the physiology for a particular experiment are included in such a medium.

There are at least two advantages to using a chemically defined medium for growing a culture of bacteria from which a secreted metabolite will be purified. The first is that this medium will be more specialized for the growth of the desired strain and therefore less likely to support the growth of contaminants. The second is that a defined medium is less likely to contain substances that may interfere with the structural characterization of the metabolite. For example, a chemically defined medium typically has several types of inorganic substances but only one type of organic substance. Complex media have many organic substances and it is the organic molecules that are most likely to interfere.

A recipe for a chemically defined medium that is known to allow growth of the desired bacterial strain, or a closely related strain, should be used. Liquid media in glass flasks are usually used unless the strain will only grow on agar medium. In addition to allowing for good growth of the strain, some other issues should be considered. The chosen medium should allow for secretion of the desired metabolite. Avoid using substances that are likely to give the medium color. If the metabolite that you plan to purify is fluorescent, do not use a medium that includes a fluorescent substance.

The starter culture and the subculture

To obtain a sufficient yield of the desired secondary metabolite, the volume of the bacterial culture from which the substance is purified should be between 0.5 and 2.0 L. The typical method for growing such a culture is as follows. A *starter culture* with a volume of 5.0–25.0 mL is inoculated with a single colony of the desired strain and then incubated until the culture is in exponential phase. A *subculture* of 0.5–2.0 L is then inoculated with a small aliquot of the starter culture. The subculture is incubated until it is in exponential phase. If one inoculates the large culture directly from a colony on a plate instead of using this two-step method, there is a chance that the liquid culture will not grow and the procedure will have to be repeated with a fresh, large-volume batch of medium.

3.2 MONITORING THE GROWTH OF BACTERIAL CULTURES BY QUANTIFICATION OF TURBIDITY

Calculation of the cellular density of a culture from spectrophotometric data

The rate of growth of a bacterial culture may be quantified by measuring the increase in *turbidity* as the number of bacteria per milliliter increases (Koch, 1970). This quantity is sometimes called the *cellular density of the culture* and is measured with a *spectrophotometer*. Most spectrophotometers provide this measurement as a percentage of *transmittance*, *T*, or as a unitless *absorbance*, *A*. Visible radiation with a wavelength of 600.0 nm is typically used. Transmittance is denoted as T_{600} and absorbance as A_{600}. The use of the terms spectrophotometer, transmittance, and absorbance in chemistry for examination of molecular substances or ions in solution is explained in a later chapter in which absorption of radiation is discussed. For chemical analysis the relationship between these two measurements is given by Eq. (3.1) (Johnson & Case, 2018).

$$A = 2.0 - \log_{10} T \qquad (3.1)$$

In microbiology, measurements of bacteria in a culture are described differently than are spectrophotometric measurements in chemistry. Although measurement of turbidity is shown as transmittance or absorbance on the display of the device, the magnitude of the absorbance is reported as the *optical density* or *OD* of the culture (Matlock, Beringer, Ash, Page, & Allen, 2011; Tortora et al., 2019). This is because visible radiation that contacts a bacterium is *scattered* so that none of it reaches the photometer (detector) of the device. In contrast, in chemical spectrophotometry only a portion of the radiation that contacts each molecule or ion is *absorbed*, while the remainder travels to the photometer. For microbiological assays Eq. (3.2) is used.

$$OD = 2.0 - \log_{10} T \qquad (3.2)$$

Although turbidity may usually be quantified as either transmittance or optical density, in some cases it may be necessary to mathematically convert transmittance data to optical density data, or *vice versa*, after completion of the experiment. Eq. (3.2) is convenient for converting transmittance to *OD*, whereas a rearrangement [Eq. (3.3)] is more convenient for converting *OD* to transmittance.

$$T = 100.0 \times 10^{-OD} \tag{3.3}$$

If a culture is monitored by optical density, the following relation may be used for quantification. A culture of *Escherichia coli* bacteria with a cellular density of 1×10^8 mL^{-1} has an OD_{600} of 0.6 (Winfrey, Rott, & Wortman, 1997). This observation may be used to derive Eq. (3.4) for calculating the density of cultures that have an OD_{600} less than or greater than 0.6.

$$\text{cellular density of culture (mL}^{-1}) = \frac{1 \times 10^8}{10^{(0.6 - OD_{600})}} \tag{3.4}$$

One may find it more convenient, however, to monitor the culture by transmittance because Eq. (3.5), which is simpler, may then be used to calculate the density.

$$\text{cellular density of culture (mL}^{-1}) = \frac{25.0\%}{T_{600}} \times 10^8 \tag{3.5}$$

A bacterium of *E. coli* has the shape of a rod that is 1.0 μm wide and 2.0 μm long (Tortora et al., 2019). These equations for calculation of the cellular density of a culture should be valid for species that are similar in size and shape to *E. coli*.

Dilution of an aliquot of the culture and correction for the dilution

Quantification of the OD_{600} of a bacterial culture is most accurate when the value reported by the spectrophotometer is between 0.1 and 1.0. This corresponds to values of T_{600} between 80.0% and 10.0%. If the OD_{600} that is measured is >1.0, and thus the T_{600} is <10.0%, it is best to repeat the measurement with a diluted aliquot of the culture. The sample of culture should be diluted into fresh culture medium.

Two issues must be considered when correcting the spectrophotometric data with the dilution factor. The first is that the dilution decreases the OD_{600} whereas it increases the T_{600}. The second is that the relationship between cellular density and OD_{600} involves an exponential function whereas the relationship between cellular density and T_{600} is linear. The two methods for correcting the data for the dilution are as follows:

- If T_{600} is measured, the value should be *divided* by the dilution factor to give the T_{600} of the culture prior to the dilution. This corrected T_{600} may then be placed in the appropriate equation.
- If OD_{600} is measured, do *not* apply the dilution factor directly to this value. Because of the exponential term in the equation that gives the cellular density from the OD_{600}, the dilution factor must be applied *after* the cellular density of the diluted sample is calculated. Calculate the cellular density of the diluted culture with the appropriate equation. To correct the calculated density of the diluted culture, *multiply* it by the dilution factor.

Calculation of the minimal generation time of a species of bacteria

Most species of bacteria reproduce by binary fission. If a culture is grown in ideal circumstances, the bacteria will reproduce as quickly as their physiology allows. Ideal circumstances include: the optimal temperature for growth, sufficient nutrients in the medium, the presence of molecular oxygen (if the strain requires it for optimal growth), and a large enough volume of medium to minimize inhibitory effects of bacterial waste. In such a case the density of the culture will increase exponentially. During a specified period of incubation the density will increase by a factor of 2^n, where n is the number of cellular fissions (generations) that occurred during that period (Tortora et al., 2019).

The time required for a bacterium to replicate its genome and divide to produce two daughter cells is the *generation time*. If the density of the culture is measured at two points in time while the culture is growing exponentially, and the number of generations calculated for this interval of time, the generation time may then be calculated. If the generation time for the strain of interest has been reported in the scientific literature, the value calculated for the strain currently being used may be compared to verify identity. It is important to consider the temperature at which the culture is being grown because rate of growth is very sensitive to temperature.

3.3 REMOVAL OF BACTERIA FROM A CULTURE BY CENTRIFUGATION AND FILTRATION

Sedimentation of the cells in a culture by centrifugation

In centrifugation, the test tubes or bottles that hold the solution of analyte are placed in a *rotor* and the center of the rotor is mounted on the *spindle*. Rotation of the spindle turns the rotor. For some centrifuges, there is more than one type of rotor that may be used. Sedimentation of cells is usually performed in a *swinging-bucket rotor*. In this type of rotor, the buckets that hold the test tubes or bottles are vertical before rotation begins, but reorient to a horizontal position after rotation starts.

The force exerted on the substances in the solution of analyte is known as the relative centrifugal force (RCF). The RCF is quantified as a multiple of the force of gravity, *g*. The equation that is used to calculate the RCF from the revolutions per minute (rpm) of the rotor may be found in the guide provided by the manufacturer of the centrifuge and rotor. The RCF is directly proportional to the distance from the center of the spindle. This distance is the *radius* and if swinging buckets are horizontal, the maximum radius is at the bottom of the bucket.

An example of the RCF in a swinging-bucket rotor that might be used for sedimentation of bacteria is as follows. If the radius to the bottom of each bucket is 20.9 cm, and it is revolving at 4200.0 rpm, the RCF at the bottom of the bucket will be 4122.0 × *g*. At least 90.0% of the bacteria in a culture will be sedimented by such a centrifugation. The supernate that includes the remaining bacteria is then decanted into a new container. Filtration may then be used to eliminate the residual bacteria.

Removal of the cells from a culture by filtration

A typical method for performing this type of filtration is as follows. A Fisherbrand circular filter that was purchased from Fisher Scientific is used. This filter is made of glass fiber, is of the G4 grade, and has a retention of 1.2 μm (a similar filter from another supplier may be used as an alternative). The filter is placed in a *Buchner funnel*, and the filter−funnel combination is placed in the neck of a *filtering flask* (Fig. 3.1). A filtering flask has a conical shape, the diameter of the opening at the top is smaller than the diameter of the base. It has a side arm (port) and is constructed of thicker glass than is found in a typical flask. This flask is used to create a vacuum underneath the

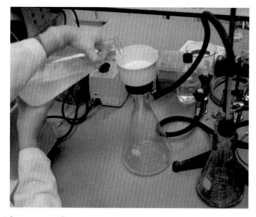

FIGURE 3.1 After centrifugation of the culture to remove the majority of the bacteria, filtration is used to remove residual bacteria from the supernate. *Reprinted with permission from Crowley, T. E., & Kyte, J. (2014).* Experiments in the purification and characterization of enzymes: A laboratory manual. *Amsterdam; Boston, MA: Elsevier/AP, Academic Press is an imprint of Elsevier. Retrieved from <https://lccn.loc.gov/2014395777.>.*

Buchner funnel. The thick glass prevents the flask from collapsing. Flexible tubing that will not collapse when a vacuum is applied to connect the side arm of the flask to an another filtering flask that functions as a *vacuum trap*. The trap is connected with tubing to a vacuum pump. The presence of the trap assures that no liquid will enter the pump.

The supernate is poured into the filter—funnel assembly and air is removed from the flasks by the pump. The vacuum drives the flow of supernate through the filter and the clarified liquid is collected in the flask that is directly below the funnel.

3.4 LIQUID–LIQUID EXTRACTION OF A METABOLITE FROM THE SUPERNATE OF A CULTURE

Two liquids are *immiscible* if they segregate into distinct layers when poured into the same container. The reason for this segregation is usually that one of the liquids is more polar than the other. The method in which a substance that has been dissolved in one solvent is transferred into another solvent is known as *solvent extraction* or *liquid–liquid extraction.*

To purify a secreted metabolite from an aqueous bacterial supernate, an organic solvent that has sufficient nonpolar character such that it is immiscible with the supernate is used. In the procedures described in this book, the organic solvent is ethyl acetate. A *separatory funnel*, which has a removable glass plug in the round opening at the top and a rotating *stopcock* in the opening at the bottom, is used. This apparatus should be made of glass or other material that is resistant to the organic solvent that will be used.

Equal volumes of the supernate and the organic solvent are poured into the separatory funnel. The glass plug is inserted into the opening at the top and the funnel is shaken to intersperse the two liquids, thus creating an *emulsion*. The funnel is then mounted vertically on a ring stand. The emulsion segregates within a few minutes into two layers, the liquid having the greater density on the bottom. If the two liquids in the funnel are an aqueous solution and ethyl acetate, the aqueous solution will be on the bottom because the density of ethyl acetate is 0.90 g mL^{-1}. The stopcock is then opened for enough time to drain the liquid in the bottom layer into a beaker or flask. The liquid that is no longer needed is discarded. More extractions may then performed on the liquid that was saved.

3.5 SOLID-PHASE EXTRACTION OF A METABOLITE FROM THE SUPERNATE OF A CULTURE

An alternative to using ethyl acetate and liquid–liquid extraction for selection of a metabolite from a bacterial supernate is adsorption onto a solid phase. This technique is most commonly known as *solid-phase extraction* or SPE. The solid phase is also known as the *adsorbent*. Although elution of the metabolite from the adsorbent usually involves the use of an organic solvent, a solvent that is miscible with water may be used. An example of such a solvent is methanol. There are two advantages to using methanol rather than ethyl acetate for purification of a metabolite. One is that methanol is the smaller of the two molecules and therefore less likely to interfere with spectroscopic or spectrophotometric assays. The other is that ethyl acetate has an offensive odor, whereas methanol does not.

An example of an adsorbent that may be used for this method is beads of Amberlite XAD-2 from MilliporeSigma. This material is a hydrophobic copolymer of styrene-divinylbenzene resin. The adsorbent is dispersed into an aqueous supernate, thus creating a *slurry*. In this slurry, the nonpolar portion of molecules of a particular metabolite may bind to the beads. The bound substances may be released by immersing the adsorbent in a solvent such as methanol that is less polar than water.

Although the solvent of the eluant will be predominantly the organic liquid, there will also be some water. It is best to minimize the amount of water because it is less volatile than most organic liquids, and therefore harder to remove in the evaporation procedure that is discussed in a later section in this chapter.

The method for minimizing water in the eluant is as follows. The mixture of aqueous supernate and adsorbent is transferred to a cylindrical tube that is positioned vertically and has a stopcock at one end (Fig. 3.2). This creates a *chromatographic column*. The method to be used in this case is not chromatography; however, the use of the column arrangement provides greater control as the solvent is changed from aqueous to organic. The excess aqueous solution is allowed to drip out of the column and then the organic solvent is poured into the top of the column. As the organic solvent drips out of the column, it carries with it the metabolite that has been released from the adsorbent.

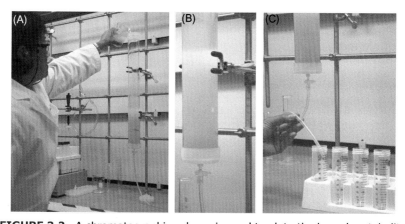

FIGURE 3.2   A chromatographic column is used to elute the bound metabolite from the adsorbent. (A) While the stopcock at the bottom of the column is closed, the aqueous slurry of adsorbent, with bound metabolite, is poured into the column. (B) The beads are allowed to settle to the bottom. (C) The stopcock is opened for enough time to allow the excess aqueous solution to drip out of the column. The aqueous solution is discarded. The organic solvent is then poured into the column and the stopcock is opened for enough time to allow this liquid to drip out of the column. This eluant contains the metabolite that has been released from the adsorbent. It is not necessary to use an automated fraction collector. *Reprinted with permission from Crowley, T. E., & Kyte, J. (2014). Experiments in the purification and characterization of enzymes: A laboratory manual. Amsterdam; Boston, MA: Elsevier/AP, Academic Press is an imprint of Elsevier. Retrieved from <https://lccn.loc.gov/2014395777.>.*

3.6 DRYING, RESUSPENSION, QUANTIFICATION OF YIELD AND STORAGE OF A PURIFIED METABOLITE

Drying

The organic solvent in which the metabolite is dissolved, after liquid–liquid extraction from the supernate of the culture or elution from the adsorbent after SPE from the supernate, may be removed by evaporation. There are two reasons for drying the preparation. The first is that the metabolite may then be resuspended in a smaller volume, providing a more concentrated solution of this substance. The second is the option of using a different solvent for resuspension than was used for extraction or elution.

Removal of water associated with an organic solvent that is not miscible with water

If the solvent that was used for extraction or elution is not miscible with water, and is less dense than water (e.g., ethyl acetate), the following will probably be observed. A small volume of water from the supernate ($\sim 1.0\%$ of the volume of the organic extract) may be present at the bottom of the container, underneath the organic liquid. Because the two liquids are immiscible, the water will appear to be a bubble trapped between the wall of the container and the organic liquid. Because all of the metabolite is expected to be in the organic liquid, and water evaporates more slowly than most of these solvents, it is best to eliminate the water before beginning the process of evaporation.

The water may be removed by adding anhydrous magnesium sulfate to the container. This inorganic salt will sink to the bottom and adsorb the water. The organic liquid, which contains the metabolite, may then be decanted into another container. This new container, as described below, should be suitable for evaporation. Evaporation of the organic solvent may then be conducted as described below.

Removal of water that is mixed with an organic solvent

If the organic liquid that was used for extraction or elution is miscible with water (e.g., methanol), then any water in the preparation cannot be removed with magnesium sulfate. The water has to be removed by the same process by which the organic solvent is removed, that is, evaporation. Water is less volatile than most organic liquids. Even if the water represents only 1.0% of the total volume, its presence will probably increase the time required for the evaporation to be completed.

Evaporation of the organic solvent

The organic liquid in which the metabolite is dissolved should be poured into a beaker. The diameter of the opening at the top of a beaker is the same as the diameter of the base. Do not use a container in which the opening at the top has a smaller diameter than the diameter of the base (e.g., an Erlenmeyer flask). Evaporation is very slow in this type of flask.

Evaporation should be conducted in a fume hood. In a hood, there is a flow of air across the open beaker and upward toward the exhaust port. This is helpful for two reasons. The first is that it prevents the workers in the laboratory from being exposed to toxic vapors. The second is that it accelerates the process of evaporation.

The dimensions of the beaker used to initiate evaporation should allow for rapid evaporation of the solvent. For the exercises in this book, the volume of the first beaker should be between 1.0 and 4.0 L. As the volume of the liquid decreases, it may sequentially be transferred to smaller beakers. A beaker with a volume of 50.0 mL is appropriate for the final step in which the last portion of liquid evaporates, leaving the residue that includes the dry metabolite.

Resuspension

If a metabolite is soluble in water, and in one or more organic solvents, there is an advantage to resuspending it in an organic liquid rather than water. Most organic liquids are more volatile than water [dimethyl sulfoxide (DMSO) is an exception as explained below]. At some time after the dry metabolite is resuspended in the solvent, it may be necessary to prepare a solution of the metabolite in a different liquid. An organic solvent may be removed by evaporation more rapidly than water may be removed. The dry metabolite may then be resuspended in the alternative solvent.

Although DMSO is suggested as the solvent for samples to be examined with nuclear magnetic resonance spectroscopy in exercises in this book, it is not recommended as the solvent for resuspension of the entire preparation of the metabolite. The boiling point of this organic liquid under atmospheric pressure is 189.0°C. If the metabolite is dissolved in DMSO, it is difficult to remove this solvent.

To determine the appropriate volume of solvent in which to resuspend the dry metabolite, consider the limit of solubility and the minimum concentration needed for the assays that will eventually be performed. An example of the typical range of concentrations is as follows. In the exercises presented in later chapters in which detailed protocols are provided for purification of metabolites, the molar masses of the substances purified will probably be between 200.0 and 600.0 g mol^{-1}. The molar yield of these molecules will probably range from 10.0 to 80.0 μm, corresponding to a mass yield of 4.0−18.0 mg. If the purified metabolites are resuspended in the recommended volumes, that is, from 2.0 to 8.0 mL, then the molar concentration of the substances will be between 2.0 and 10.0 mM, corresponding to a mass per volume concentration of 1.0−2.0 mg mL^{-1}.

Quantification of yield

Assay of the absorbance of ultraviolet (UV) or visible radiation by a solution of the purified metabolite followed by some calculations will reveal the concentration of this substance in the resuspended sample that was described above. This calculation is explained in the later chapter that focuses on absorption of radiation. The concentration of the metabolite, and the volume in which the substance was resuspended, may then be used to calculate the yield.

Storage of a purified metabolite

If the metabolite is dissolved in an organic solvent, the test tubes used for storage must be constructed of material that is resistant to the solvent. The resuspended sample should be divided into several portions and each stored in a distinct test tube. If the preparation is divided in this way, only a portion of it is exposed to the atmosphere when a aliquot is removed for an experiment. To prevent evaporation of the solvent, the caps should be tightly sealed. To prevent photolysis of the metabolite, the test tubes should be shielded from UV and visible radiation. If a refrigerator or freezer is available, storage at the lower temperature will minimize the possibility of microbial growth in the samples. Although some metabolites may be stable at the typical ambient temperature in a laboratory

(22.0°C), storage at the lower temperature may also prevent degradation of the purified substance.

BIBLIOGRAPHY

Atlas, R. M. (2010). *Handbook of microbiological media* (4th ed.). Washington, DC; Boca Raton, FL: ASM Press; CRC Press/Taylor & Francis. Retrieved from <https://lccn.loc.gov/2009047096>.

Crowley, T. E., & Kyte, J. (2014). *Experiments in the purification and characterization of enzymes: A laboratory manual*. Amsterdam; Boston, MA: Elsevier/AP, Academic Press is an imprint of Elsevier. Retrieved from <https://lccn.loc.gov/2014395777>.

Holt, J. G., Krieg, N. R., Sneath, P. H. A., Staley, J. T., & Williams, S. T. (1994). *Bergey's manual of determinative bacteriology* (9th ed.). Baltimore, MD: Wolters Kluwer/Lippincott Williams & Wilkins. Retrieved from <https://shop.lww.com/Bergey-s-Manual-of-Determinative-Bacteriology/p/9780683006032>.

Johnson, T. R., & Case, C. L. (2018). *Laboratory experiments in microbiology* (12th ed.). Hoboken: Pearson. Retrieved from <https://lccn.loc.gov/2017039734>.

Koch, A. L. (1970). Turbidity measurements of bacterial cultures in some available commercial instruments. *Analytical Biochemistry*, *38*, 252–259. <https://www.ncbi.nlm.nih.gov/pubmed/4920662>.

List_of_Prokaryotic_Names_with_Standing_in_Nomenclature. (2018). *Microbial knowledge base (LPSN)*. Sudbury, MA. Retrieved from <http://www.bacterio.net>.

Matlock, B. C., Beringer, R. W., Ash, D. L., Page, A. F., and Allen, M. W. (2011). *Differences in bacterial optical density measurements between spectrophotometers (technical note 52236)*. Thermo Fisher Scientific. Retrieved from <https://www.thermofisher.com/search/supportSearch?query = &navId = 4294959596&refinementQuery = tn52236>.

Strain_Info. (2018). *Microbial knowledge base*. Ghent, Belgium: Ghent University. Retrieved from <http://www.straininfo.net>.

Tortora, G. J., Funke, B. R., & Case, C. L. (2019). *Microbiology: An introduction* (13th ed.). Boston, MA: Pearson. Retrieved from <https://lccn.loc.gov/2017044147>.

Whitman, W. B. (Ed.), (2015). *Bergey's manual of systematics of archaea and bacteria*. Wiley Online Library: John Wiley & Sons, Inc. in association with Bergey's Manual Trust. Retrieved from <http://onlinelibrary.wiley.com/book/10.1002/9781118960608>.

Winfrey, M. R., Rott, M. A., & Wortman, A. T. (1997). *Unraveling DNA: Molecular biology for the laboratory*. Upper Saddle River, NJ: Prentice-Hall. Retrieved from <https://lccn.loc.gov/96036610>.

Chapter 4

Absorption of ultraviolet, visible, and infrared radiation

4.1 ABSORPTION OF ULTRAVIOLET AND VISIBLE RADIATION BY ORGANIC MOLECULES AND IONIZED METALS IN SOLUTION

The visible region of the spectrum of electromagnetic radiation is typically considered to be comprised of wavelengths from 400 nm (violet light) to between 700 and 800 nm (red light). Ultraviolet (UV) radiation comprises shorter wavelengths. The wavelengths of *near UV* radiation are between 200 and 400 nm, whereas *far UV* radiation comprises wavelengths between 50 and 200 nm (Banwell & McCash, 1994; Vollhardt & Schore, 2014). Far UV is also known as *vacuum UV* because these wavelengths travel through a vacuum much more efficiently than they do through atmospheric gases.

Most organic molecules that are not associated with a metallic ion lack color because they do not absorb a significant amount of visible radiation. Absorption of UV radiation is, however, common in organic molecules (Crews, Rodríguez, & Jaspars, 2010; Field, Sternhell, & Kalman, 2013; Vollhardt & Schore, 2014). Absorption of the near UV is the most useful for analytical chemistry because the devices typically used for the measurement do not contain the vacuum chamber that is necessary for propagating far UV. Absorption of the near UV radiation is usually due to the presence of a carbon–oxygen double bond, a carbon–carbon triple bond, or a series of conjugated carbon–carbon double bonds. Each UV-absorbing structure within an organic molecule is known as a *UV chromophore* (Crews et al., 2010; Field et al., 2013; Taber, 2007).

Color may be observed in an aqueous solution containing ions of a transition metal (Atkins, Jones, & Laverman, 2016). Examples include titanium (Ti), iron (Fe), nickel (Ni), and

Purification and Characterization of Secondary Metabolites.
DOI: https://doi.org/10.1016/B978-0-12-813942-4.00004-8
© 2020 Elsevier Inc. All rights reserved.

copper (Cu) that are in the d-block of the periodic table. Ions of iron(III) absorb UV radiation and this may also be true of the other three metals. The colors, however, are due to the absorption of visible radiation. The color produced by titanium (III) is violet, iron(III) is yellow, nickel(II) is green, and copper (II) is blue.

In physical and analytical chemistry, the spectrum of absorption of a substance is typically obtained from a solution of the pure substance. The substance is referred to as the *analyte*. If the spectrum includes both the UV and visible wavelengths, it is referred to as a *UV-vis spectrum*. Such spectra are sometimes described as *electronic spectra* because of the shift of electrons to new orbitals (refer to definition of *absorption* below).

Definitions for the use of UV and visible radiation on analytes in solution

Absorption

Retention of the energy in incident radiation of a particular wavelength by the analyte. If the incident radiation is UV or visible, this retention of energy is due to electrons in the molecules of analyte shifting to new orbitals in higher energy levels. It is a theoretical property of a molecule of analyte rather than an experimentally measured value.

Transmittance

An experimental measurement of the percentage of incident radiation of a particular wavelength that exits the chamber holding the solution of analyte. Values can range from 0.0% to 100.0%. In a mathematical equation it is denoted as an uppercase letter in oblique font with the wavelength in nanometers as a subscript. Transmittance of UV radiation with a wavelength of 312 nm is therefore T_{312}.

Absorbance

An experimental measurement of the amount of incident radiation of a particular wavelength that is absorbed by the solution of analyte. Absorbance is a numerical value that may be zero or a positive number and it does not have a unit associated with it. The upper limit depends on the device being used and possibly other factors. In a mathematical equation it is denoted as an uppercase letter in oblique font with a subscript indicating the wavelength in nanometers. Absorbance of UV radiation with a wavelength of 312.0 nm is therefore A_{312}. It has an inverse, logarithmic relation to transmittance (Johnson & Case, 2018). For radiation with a wavelength of 312.0 nm, Eq. (4.1) shows the relationship between absorbance and transmittance.

$$A_{312} = 2 - \log_{10}(T_{312}) \tag{4.1}$$

It is the value referred to in discussions of numerical data and it is plotted on the *y*-axis in a spectrum.

Spectroscopy

Examination of the variation in absorption of an analyte over a range of wavelengths of incident radiation. The range is generated by a *scan of wavelengths*, and the data are displayed as a spectrum. The typical goal is to gain insight into the structure of the analyte molecule.

Spectrophotometry

Quantification of the absorbance of a solution of an analyte at a single wavelength, or a few wavelengths, of incident radiation (Atkins et al., 2016; Crowley & Kyte, 2014). In this case the device is usually set to a *fixed wavelength* for each measurement. The data that are recorded and a calculation with the *Beer−Lambert law* (discussed below) are then used to determine the molar concentration of a substance in the solution.

Spectrophotometer

This device consists of a *spectrometer coupled to a photometer.* The spectrometer contains a source of radiation and components that allow only a single wavelength of radiation to project into the solution of analyte at any given moment. In spectroscopy the wavelength of this incident radiation is gradually changed to create a scan. In spectrophotometry this wavelength is usually fixed for each measurement. The photometer is also known as the *detector,* and it quantifies the intensity of the radiation that exits from the chamber holding the solution of analyte. This signal is then converted to a value of transmittance or absorbance (defined above). When comparing a variety of analytes, those absorbing more radiation will show lower transmittance.

Examples of the use of UV and visible radiation on analytes in solution

Examples of the following three applications are provided in this chapter. The first is an example of spectroscopy. The second and third are examples of spectrophotometry and the use of the Beer–Lambert law.

1. Our knowledge of the wavelength of maximum absorption, and relative intensity of absorption, of various UV chromophores may be used to gain insight into the structure of an organic analyte.
2. The concentration of the organic analyte in solution may be quantified by measuring the absorption at a single wavelength.
3. Interaction between an organic analyte and an ionized metal may be detected by the change in the magnitude of absorption at a single wavelength that results from binding.

The Beer–Lambert law

For this calculation a *molar extinction coefficient* for the analyte at the specified wavelength must be available. In a mathematical equation this physical constant is represented by the Greek letter ε (epsilon) followed by the wavelength (in nanometers) as a subscript. This constant is the theoretical absorption of an aliquot of a 1.0 M solution of the analyte in a cuvette with a pathlength of 1.0 cm. The unit is $(molar)^{-1}$ $(centimeter)^{-1}$. The concentration of analyte is much less than 1.0 M in the examples and exercises in this book, therefore ε will be expressed in $(millimolar)^{-1}$ $(centimeter)^{-1}$. For example, quantification of the concentration of desferric enterobactin by means of its absorption of UV radiation with a wavelength of 312 nm (A_{312}) is calculated using $\varepsilon_{312} = 9.4$ mM^{-1} cm^{-1} (Young & Gibson, 1979).

If A represents the absorption of a solution of a pure substance at a specified wavelength, ε is the molar extinction coefficient in $(millimolar)^{-1}$ $(centimeter)^{-1}$ for that substance at that wavelength, c is the concentration of the substance in millimolar, and l is the pathlength in centimeters, then the Beer–Lambert law (Atkins et al., 2016; Crews et al., 2010; Field et al., 2013) is represented by the following equation:

$$A = \varepsilon c l \qquad (4.2)$$

4.2 RECORDING THE SPECTRA OF ABSORPTION OF ULTRAVIOLET AND VISIBLE RADIATION BY ORGANIC MOLECULES AND IONIZED METALS IN SOLUTION

Binding of a metallic ion by an organic compound is known as *chelation*. In the organometallic compound that is formed, the organic portion is known as the *ligand*. Recording the UV-vis spectra of solutions of metallic ions, organic compounds, and organometallic compounds provides insight into the process of chelation.

Enterobactin is one of the best characterized siderophores. It is expressed by the enteric bacterium *Escherichia coli* and some other bacterial species (Pollack & Neilands, 1970; Raymond & Carrano, 1979; Raymond, Dertz, & Kim, 2003). An aqueous solution of iron(III) that is not associated with an organic ligand absorbs violet light that is wavelengths from 400 to 430 nm. This makes the solution appear yellow. Addition of enterobactin to such a solution changes the color to brown (shown earlier in this book), or to red or purple (Pecoraro, Harris, Wong, Carrano, & Raymond, 1983; Young & Gibson, 1979). This is presumably due to chelation of the ions by this siderophore. The variation in the color of ferric enterobactin is due to variation in the pH of the solution. Ions of iron(III) free of ligand and ferric enterobactin both absorb a range of wavelengths; however, the change in color is presumably due to the change in the wavelength of maximum absorption (λ_{max}) that occurs when the ligand binds the ion. This provides a spectrophotometric method for detection of the interaction between ion and ligand.

Dry enterobactin was obtained from a commercial supplier (Sigma-Aldrich_Co_enterobactin, 2017) and resuspended in acetonitrile at a concentration of 1.0 mg mL^{-1} (1.5 mM). Crystals of iron(III) chloride were dissolved in deionized water at a concentration of 10.0 mg mL^{-1} (60.0 mM). The three samples examined were the following:

1. Enterobactin (13.0 μg in 13.0 μL of acetonitrile);
2. FeCl$_3$ (aq) (10.0 μg in 1.0 μL of deionized water);
3. A mixture of 13.0 μg enterobactin in 13.0 μL of acetonitrile and 10.0 μg of FeCl$_3$ in 1.0 μL of deionized water.

To record the UV-vis spectrum of these samples, each was diluted with deionized water to a final volume of 500.0 μL.

This generated three diluted samples with the following concentrations:

1. Enterobactin at a concentration of 40.0 μM (20.0 nmol in 500.0 μL of 2.6% acetonitrile);
2. Iron(III) at a concentration of 120.0 μM (60.0 nmol in 500.0 μL of deionized water);
3. A mixture 40.0 μM enterobactin and 120.0 μM iron(III) in 500.0 μL of 2.6% acetonitrile.

Note: The molar ratio of enterobactin:iron(III) in the third sample was 1:3. Acetonitrile at a concentration of 2.6% does not interfere with interpretation of these UV-vis spectra.

The diluted samples were then assayed in a quartz cuvette with a pathlength of 1.0 cm. Wavelengths from 190.0 to 1100.0 nm, which include the near UV and the visible region, were assayed.

4.3 EVALUATING THE SPECTRA OF ABSORPTION OF ULTRAVIOLET AND VISIBLE RADIATION BY ORGANIC MOLECULES AND IONIZED METALS IN SOLUTION

Evaluation of the spectra as spectroscopy

The UV region of the spectrum of desferric enterobactin shows peaks at 210.0, 245.0, and 312.0 nm (Fig. 4.1). This profile of three peaks is similar to the reported spectrum of

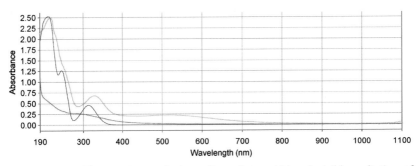

FIGURE 4.1 The spectrum of absorption of near UV and visible radiation of iron(III), enterobactin by itself and a mixture of enterobactin and iron(III), displayed at low resolution. The spectrum of iron(III) is in red, enterobactin by itself is in black, and the mixture is in green. This low-resolution image shows that the three peaks of absorption in the UV region that are characteristic of desferric enterobactin and are only slightly altered by addition of iron(III) to a solution of this siderophore.

acetophenone (Field et al., 2013; Gutsche & Pasto, 1975). Acetophenone is an acetyl group linked to a phenyl group; therefore, it includes a carbonyl adjacent to a benzene ring. Both of these structures are known to be UV chromophores. Enterobactin has three benzene rings, each of which is linked to a carbonyl. This part of the enterobactin structure is probably the reason for the similarity in the UV spectra of these two molecules.

The UV-vis spectrum of unchelated ions of iron(III) shows strong absorption at the shortest wavelength that was assayed, 190.0 nm, and the absorption gradually decreases as the wavelength increases. The absorption from 400.0 to 430.0 nm causes the solution to appear yellow.

Evaluation of the spectra as spectrophotometry

To verify that the absorption quantified in these spectra is due to enterobactin rather than a contaminant, and to determine if chelation occurred in the mixture, the published values for the molar extinction coefficients may be used. Desferric enterobactin absorbs UV radiation and the ε_{312} has been reported to be 9.4 mM^{-1} cm^{-1} (Young & Gibson, 1979). Ferric enterobactin absorbs both visible and UV radiation and the ε_{495} has been reported to be 5.6 mM^{-1} cm^{-1} (Harris et al., 1979).

Desferric enterobactin

The A_{312} recorded here for desferric enterobactin (Fig. 4.1), and the published ε_{312} for desferric enterobactin that was mentioned earlier, may be used to calculate the concentration of enterobactin when it was dissolved in acetonitrile, that is, before addition of iron(III) and dilution. The length of the path of light passing through the cuvette was 1.0 cm. If [desferric enterobactin] represents the millimolar concentration of this molecule in the cuvette, that is, after dilution, then application of the Beer–Lambert law gives the following equation:

$$A_{312} = \varepsilon_{312} \times [\text{desferric enterobactin}] \times 1.0\text{cm} \qquad (4.3)$$

Inserting 0.450 for A_{312} and the value of ε_{312} mentioned above, the postdilution concentration of desferric enterobactin is calculated to be 0.0479 mM.

The predilution concentration of desferric enterobactin is calculated with the following equation:

$$0.0479 \text{mM} \times \frac{500.0 \mu\text{L}}{13.0 \mu\text{L}} = 1.84 \text{mM} \qquad (4.4)$$

Although not a precise match, this value is close enough to the expected value of 1.50 mM to suggest that the A_{312} was due to absorption by desferric enterobactin rather than a contaminant.

Ferric enterobactin

In the region of UV radiation, the spectrum for the mixture of $FeCl_3$ (aq) and enterobactin is similar to that of enterobactin by itself; however, a new peak appears at 510.0 nm in the visible region in the mixture. The new peak suggests that enterobactin has chelated iron(III) and is now ferric enterobactin (Fig. 4.2).

The A_{510} recorded here for what is presumed to be ferric enterobactin, and the published ε_{495} for ferric enterobactin that was mentioned earlier, may be used to calculate the concentration of enterobactin when it was dissolved in acetonitrile. The length of the path of light passing through the cuvette was 1.0 cm. If [ferric enterobactin] represents the millimolar concentration of this molecule in the cuvette, that is, after dilution, then application of the Beer−Lambert law gives the following equation:

$$A_{510} = \varepsilon_{495} \times [\text{ferric enterobactin}] \times 1.0 \text{cm} \qquad (4.5)$$

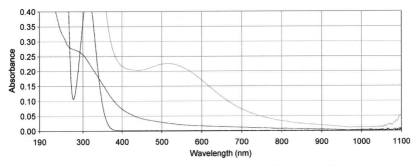

FIGURE 4.2 The spectrum of absorption of near UV and visible radiation of iron(III), enterobactin by itself and a mixture of enterobactin and iron(III), displayed at high resolution. The spectrum of iron(III) is in red, enterobactin by itself is in black, and the mixture is in green. This high-resolution image shows that the absorption of visible radiation with a wavelength of 510.0 nm is not very significant for either iron(III) or enterobactin assayed independently; however, the mixture of these two substances generates a strong peak.

Inserting 0.225 for A_{510} and the value of ε_{495} mentioned earlier, the postdilution concentration of the presumed ferric enterobactin is calculated to be 0.0402 mM.

The predilution concentration of the presumed ferric enterobactin is calculated with the following equation:

$$0.0402\text{mM} \times \frac{500.0\mu\text{L}}{13.0\mu\text{L}} = 1.55\text{mM} \qquad (4.6)$$

Although not a precise match, this value is close enough to the expected value of 1.50 mM to suggest that the A_{510} was due to absorption by ferric enterobactin. This confirms that chelation did occur in the mixture.

4.4 ABSORPTION OF INFRARED RADIATION BY ORGANIC MOLECULES

The structure within an organic molecule that absorbs a particular frequency of infrared (IR) radiation is known as an *IR chromophore* (Field et al., 2013). Experiments involving the absorption of IR radiation are usually performed to elucidate the structure of the *analyte*, that is in this case an organic molecule. This technique is not frequently used to quantify the concentration of analyte in solution or to detect interaction with another molecule; therefore, it is known as *IR spectroscopy* rather than IR spectrophotometry (for definitions of these two terms, refer to the earlier section: Definitions for the use of uv and visible radiation on analytes in solution).

The analyte may be a pure solid or a pure substance dissolved in a solvent. The device used for IR spectroscopy, like the device used for quantifying absorbance of UV and visible radiation, consists of components that project radiation of the desired frequencies into the analyte and a detector that quantifies the radiation that exits the analyte. The apparatus, however, is known as an *IR spectrometer* rather than an IR spectrophotometer (for the definition of spectrophotometer, refer to the earlier section: Definitions for the use of uv and visible radiation on analytes in solution).

The absorption that is measured by an IR spectrometer is displayed in an *IR spectrum*. The spectrum is a plot of the intensity of the absorption as a function of the *wavenumber* in cm^{-1}. The wavenumber is directly proportional to the frequency of the radiation to which the analyte is exposed (Banwell & McCash, 1994; Crews et al., 2010; Gutsche & Pasto, 1975; Vollhardt & Schore, 2014).

The format of a spectrum of the absorption of IR radiation by an analyte

There are two ways to present a spectrum that quantifies the absorption of a range of IR frequencies by an analyte.

1. The signal may be plotted as *transmittance*. This is the amount of the radiation projected toward the analyte (incident radiation) that exits the analyte and reaches the detector. Transmittance can range from 0.0% to 100.0%. A transmittance of 0.0% means that all of the radiation was absorbed, whereas 100.0% means none was absorbed. On the spectrum, 100.0% transmittance is at the top of the *y*-axis; therefore, the stronger the absorption by the analyte, the greater the *downward* deflection on the spectrum. Absorption is therefore indicated by "valleys."
2. The signal may be plotted as *absorbance*. This is the amount of the radiation projected toward the analyte (incident radiation) that is absorbed by the analyte, and therefore does *not* reach the detector. Absorbance is either zero or a positive number and there is no unit associated with it. On the spectrum, the value of zero absorbance is at the bottom of the *y*-axis. The stronger the absorption by the analyte, the greater the *upward* deflection on the spectrum. Absorption is therefore indicated by "peaks."

In a discussion of the spectrum of absorption of IR radiation, each signal is referred to as an *IR band*, rather than a valley or a peak.

4.5   RECORDING THE SPECTRA OF ABSORPTION OF INFRARED RADIATION BY ORGANIC MOLECULES

Dimethyl sulfoxide

Dimethyl sulfoxide (DMSO) is an organic solvent in which some secondary metabolites may be dissolved. Unlike most other organic solvents, DMSO does not evaporate rapidly at ambient temperature. This is convenient for analytical techniques such as nuclear magnetic resonance spectroscopy in which the analyte must be in the liquid phase. IR spectroscopy, however, is often performed on a sample in which the solvent has been allowed to evaporate. Although it is best to dissolve the metabolite of interest in a solvent that is volatile at ambient temperature, there may be metabolites for which DMSO is the only practical solvent. To properly interpret IR data for a metabolite in DMSO, a spectrum of the solvent without metabolite must also be recorded. A data-analysis program may then be used to subtract the spectrum of the solvent from the spectrum of the metabolite dissolved in the solvent.

The spectrum of absorption of IR radiation of pure DMSO was examined to provide a reference for experiments in which the analyte is dissolved in this solvent. A sample of 2.0 μL of pure DMSO liquid was applied to the sample-loading spot on the spectrometer and the spectrum was recorded (Fig. 4.3). Evaluation of this spectrum is presented in the next section.

Enterobactin

Enterobactin was one of the first siderophores to be thoroughly characterized. It is expressed by the bacterium *E. coli* and some other species of bacteria (Pollack & Neilands, 1970; Raymond & Carrano, 1979; Raymond et al., 2003).

Enterobactin was obtained from a commercial supplier (Sigma-Aldrich_Co_enterobactin, 2017) and dissolved in acetonitrile. An aliquot containing 3.0 nmol of this metabolite was applied to the sample-loading spot on the spectrometer and the solvent was allowed to evaporate. The spectrum of absorption of IR radiation of the dry enterobactin was recorded (Fig. 4.4). Evaluation of this spectrum is presented in the next section.

4.6   ELUCIDATING THE STRUCTURE OF ORGANIC MOLECULES BY EXAMINING THEIR SPECTRA OF ABSORPTION OF INFRARED RADIATION

The IR spectra of DMSO and enterobactin that are presented in the previous section may be evaluated as follows:

Dimethyl sulfoxide

Assignments of the bands of absorption of IR radiation to IR chromophores within DMSO are summarized in Table 4.1. The thionyl (S = O) is the most unique structure in DMSO; it is not very common in organic molecules. The thionyl is a good indicator for DMSO in an IR spectrum because the stretch and contraction of this bond creates the strongest band for this molecule, with a wavenumber of $1042 \, cm^{-1}$. This is a significantly

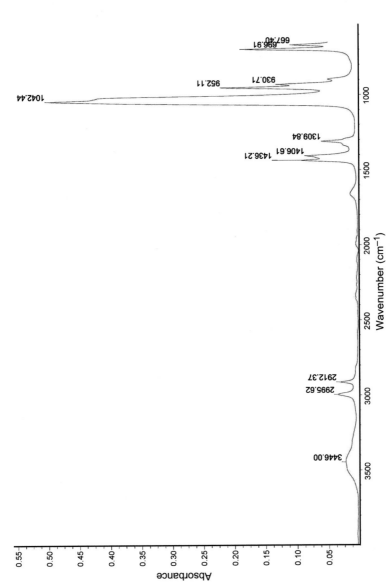

FIGURE 4.3 The spectrum of infrared radiation absorbed by dimethyl sulfoxide, an organic solvent. Assignments of the various bands of absorption to chromophores within dimethyl sulfoxide are in Table 4.1.

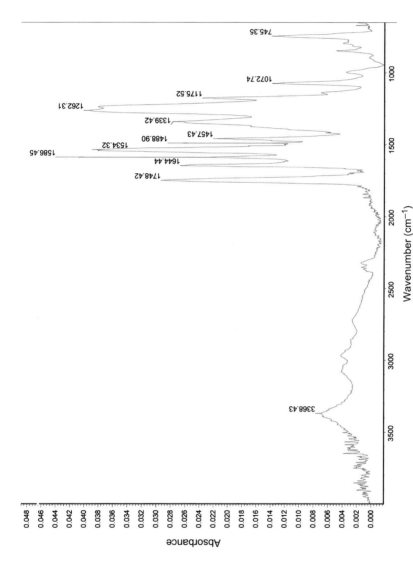

FIGURE 4.4 The spectrum of absorption of infrared radiation of enterobactin, a siderophore. Assignments of the various bands of absorption to chromophores within enterobactin are in Table 4.2.

TABLE 4.1 Dimethyl sulfoxide: correlation of bands of absorption of infrared radiation with chromophores.

Wavenumber (cm^{-1})	Absorbance	IR chromophore[a]	Vibration
696	0.180	C–S	Stretch or bend
952	0.213	C–S	Stretch or bend
1042	0.495	S = O	Stretch
1406	0.080	C–H	Bend
1436	0.090	C–H	Bend
2912	0.030	C–H	Stretch
2995	0.038	C–H	Stretch

[a]Assignments of IR chromophores based on reference tables in textbooks (Banwell & McCash, 1994; Crews et al., 2010; Field et al., 2013; Gutsche & Pasto, 1975).

TABLE 4.2 Enterobactin: correlation of bands of absorption of infrared radiation with chromophores.

Wavenumber (cm^{-1})	Absorbance	IR chromophore[a]	Vibration
745	0.013	C–H (phenyl)	Bend
1072	0.013	C–C, C–O, C–N	Stretch
1175	0.023	C–C, C–O, C–N	Stretch
1262	0.039	C–H	Bend
1339	0.027	C–C (phenyl)	Stretch
1457	0.022	C–C (phenyl)	Stretch
1488	0.014	C–H (methylene)	Bend
1534	0.038	C–C (phenyl)	Stretch
		C = O (amide)	Stretch
1586	0.022	C–C (phenyl)	Stretch
1644	0.026	C = O (amide)	Stretch
1748	0.029	C = O (ester)	Stretch
3368	0.007	C–H (phenyl) O–H (hydroxyl) N–H (amide)	Stretch

[a]Assignments of IR chromophores based on published data for enterobactin (Harris et al., 1979; Pecoraro et al., 1983) and reference tables in textbooks (Banwell & McCash, 1994; Crews et al., 2010; Field et al., 2013; Gutsche & Pasto, 1975).

lower value than the wavenumbers of the bands generated by the carbonyls ($C = O$) in enterobactin as discussed below. This is due to the fact that an atom of sulfur has a greater mass than an atom of carbon.

Enterobactin

Assignments of the bands of absorption of IR radiation to IR chromophores within enterobactin are summarized in Table 4.2. The greatest wavenumber, $3368\,cm^{-1}$, is due to stretch of C−H, O−H, and N−H bonds. The stretch of carbonyl bonds generates bands with wavenumbers of 1534, 1644, and $1748\,cm^{-1}$. The bending of various C−H bonds has a lower frequency than the stretches just mentioned, resulting in bands with wavenumbers of 1488, 1262, and $745\,cm^{-1}$. These data provide an example of how this technique may be used to verify the identity of a pure sample of a secondary metabolite.

BIBLIOGRAPHY

Atkins, P. W., Jones, L., & Laverman, L. (2016). *Chemical principles: The quest for insight* (7th ed.). New York, NY: W.H. Freeman: Macmillan Learning. Available from https://lccn.loc.gov/2015951706.

Banwell, C. N., & McCash, E. M. (1994). *Fundamentals of molecular spectroscopy* (4th ed.). London; New York, NY: McGraw-Hill. Available from https://lccn.loc.gov/94000869.

Crews, P., Rodríguez, J., & Jaspars, M. (2010). *Organic structure analysis* (2nd ed.). New York, NY: Oxford University Press. Available from https://lccn.loc.gov/2009018383.

Crowley, T. E., & Kyte, J. (2014). *Experiments in the purification and characterization of enzymes: A laboratory manual*. Amsterdam; Boston, MA: Elsevier/AP, Academic Press is an imprint of Elsevier. Available from https://lccn.loc.gov/2014395777.

Field, L. D., Sternhell, S., & Kalman, J. R. (2013). *Organic structures from spectra* (5th ed.). Chichester, West Sussex: Wiley. Available from https://lccn.loc.gov/2012046033.

Gutsche, C. D., & Pasto, D. J. (1975). *Fundamentals of organic chemistry*. Englewood Cliffs, NJ: Prentice-Hall. Available from https://lccn.loc.gov/74009714.

Harris, W. R., Carrano, C. J., Cooper, S. R., Sofen, S. R., Avdeef, A. E., McArdle, J. V., & Raymond, K. N. (1979). Coordination chemistry of microbial iron transport compounds. 19. Stability constants and electrochemical behavior of ferric enterobactin and model complexes. *Journal of the American Chemical Society*, *101*, 6097−6104. Available from https://doi.org/10.1021/ja00514a037.

Johnson, T. R., & Case, C. L. (2018). *Laboratory experiments in microbiology* (12th ed.). Hoboken: Pearson. Available from https://lccn.loc.gov/2017039734.

Pecoraro, V. L., Harris, W. R., Wong, G. B., Carrano, C. J., & Raymond, K. N. (1983). Coordination chemistry of microbial iron transport compounds. 23. Fourier transform infrared spectroscopy of ferric catechoylamide analogues of enterobactin. *Journal of the American Chemical Society*, *105*, 4623–4633. Available from https://doi.org/10.1021/ja00352a018.

Pollack, J. R., & Neilands, J. B. (1970). Enterobactin, an iron transport compound from *Salmonella typhimurium*. *Biochemical and Biophysical Research Communications*, *38*, 989–992. Available from http://www.sciencedirect.com/science/article/pii/0006291X70908193.

Raymond, K. N., & Carrano, C. J. (1979). Coordination chemistry and microbial iron transport. *Accounts of Chemical Research*, *12*, 183–190. Available from https://doi.org/10.1021/ar50137a004.

Raymond, K. N., Dertz, E. A., & Kim, S. S. (2003). Enterobactin: An archetype for microbial iron transport. *Proceedings of the National Academy of Sciences*, *100*, 3584–3588. Available from http://www.pnas.org/content/100/7/3584.abstract.

Sigma-Aldrich_Co_enterobactin. (2017). *Enterobactin from* Escherichia coli, *item E3910*. St. Louis, MO. http://www.sigmaaldrich.com/catalog/product/sigma/e3910?lang = en&region = US.

Taber, D. F. (2007). *Organic spectroscopic structure determination: A problem-based learning approach*. New York, NY: Oxford University Press. Available from https://lccn.loc.gov/2006035525.

Vollhardt, K. P. C., & Schore, N. E. (2014). *Organic chemistry: Structure and function* (7th ed.). New York, NY: W.H. Freeman and Company. Available from https://lccn.loc.gov/2013948560.

Young, I. G., & Gibson, F. (1979). Isolation of enterochelin from *Escherichia coli*. *Methods in Enzymology*, *56*, 394–398. Available from http://www.sciencedirect.com/science/article/pii/0076687979560376.

Chapter 5

High-performance liquid chromatography

5.1 THEORY AND PRACTICE OF HIGH-PERFORMANCE LIQUID CHROMATOGRAPHY

Note: The literary references provided in this section include diagrams that clarify the concepts discussed here.

The history and variations of high-performance liquid chromatography

When this technique was invented it was called high-*pressure* liquid chromatography and the abbreviation HPLC was used (Khopkar, 2009; Meyer, 2010; Xu, 2013). As the manufacturers of the devices for HPLC refined the technology, they changed the name to high-*performance* liquid chromatography. No change in the abbreviation was necessary, so "HPLC" was still used. Although much of the work done with this technique is still considered to be "high-performance" liquid chromatography, in the year 2004 the technology advanced to *ultra*-high-performance liquid chromatography and a new abbreviation, UHPLC, was introduced. In a later chapter, exercises in which either HPLC or UHPLC may be used to assay for chelation of iron by a siderophore are presented. In the next section in this chapter, an example of the use of UHPLC, coupled with detection by absorption of ultraviolet or visible radiation, is described.

HPLC may be used as a *preparative* or an *analytical* technique. The preparative variation is for purification of a unique chemical compound from a mixture of compounds. In this case the mass of material loaded into the chromatographic column is greater than that loaded in the analytical variation. As the *eluent* (solution of the loaded compounds) exits the column, *chromatographic fractions* are collected in test tubes. The various substances in the mixture may be separated into distinct

Purification and Characterization of Secondary Metabolites.
DOI: https://doi.org/10.1016/B978-0-12-813942-4.00005-X

fractions and are thereby purified. In the analytical method the eluent is not collected. In this case the liquid is monitored by a detector that is incorporated into the HPLC device, or directed into another device such as a *mass spectrometer*. The purpose of analytical HPLC is to characterize the structural and physical properties of the chemical compounds. A typical column that is used for preparative HPLC is larger than those used for analytical HPLC. The following discussion pertains to analytical HPLC.

Another variation in the use of HPLC is that it can be *normal phase* or *reversed phase*. In normal-phase HPLC the chemical structures linked to the solid support within the column are more polar than the solvent. In reversed-phase HPLC the solvent is more polar than the chemical structures linked to the solid support. This chapter focuses on reversed-phase HPLC.

Application of high-performance liquid chromatography to organic metabolites

A mixture of organic metabolites being examined with chromatography is often referred to as the *analyte*. The separation, identification, and quantification of the substances in this type of analyte must be performed with techniques distinct from those that are used to characterize biological macromolecules. Because metabolites are typically less than $2000.0 \, \mathrm{g \, mol^{-1}}$, electrophoresis in gels and molecular exclusion chromatography are not very effective for resolving them. The organic molecular matrix inside of a gel of polyacrylamide or agarose will resolve many types of polypeptides and nucleic acids but will not resolve an analyte consisting of a mixture of metabolites.

HPLC is very different than molecular exclusion chromatography. An apparatus for the molecular exclusion technique includes a vertically mounted cylinder that has a diameter of several centimeters and a length of $\sim 1.0 \, \mathrm{m}$. The cylinder is filled with porous beads of plastic that are immersed in liquid. The macromolecules in an analyte that is loaded into the top of the column are caused to migrate toward the bottom by gravity. A mechanical pump may be used to supply solvent into the top of the column; however, migration of analyte is only driven by gravity. If an analyte consisting of a mixture of polypeptides that vary in their number of amino acids is loaded into this type of column, those with fewer amino acids will enter the pores of the beads more easily than those with more amino acids. Because they are more likely to weave through the interior of the beads, the smaller polypeptides travel more slowly than the larger polypeptides.

A column for HPLC is much smaller than that used for molecular exclusion chromatography (Khopkar, 2009; Meyer, 2010; Xu, 2013). The interior diameter is between 2.0 and 5.0 mm and the length is between 50.0 and 250.0 mm. The liquid solvent that is known as the *mobile phase* for this technique is pumped through the column creating the "high pressure" that is described later. Because the movement of the mobile phase is driven by a mechanical pump rather than gravity, the column may be in either a vertical or horizontal orientation. The material inside of the column with which the molecules of analyte interact is known as the *stationary phase* and is typically unbranched alkanes linked to spherical particles.

The amount of time from when an analyte enters the HPLC column until a particular type of molecule in the analyte exits the column is known as the *time of retention* of that molecule. In HPLC, the molecules of analyte do not weave through a matrix as in electrophoresis and molecular exclusion chromatography. Because the alkanes mentioned earlier are nonpolar, hydrophobic interaction between molecules of analyte and the stationary phase decreases the rate of migration. The stronger the hydrophobic interaction, the slower the rate. An increase in the amount of nonpolar structure in an organic molecule will increase the association with the stationary phase, whereas an increase in the polarity of the molecule will decrease this association. The time of retention depends on both the size and the polarity of each type of organic molecule.

The stationary phase, the mobile phase and the chromatographic parameters

The mobile phase for HPLC must allow for sufficient solubility of the substances in the analyte so that they may be detected as they exit the column. After being purified, some metabolites are resuspended in an organic solvent because they are more soluble in this than in water and it is best to have a solution of the substance at as high a concentration as possible. Even if a particular metabolite is more soluble in an organic solvent than it is in water, it may remain soluble after being diluted into water to the concentration that is appropriate for HPLC.

Two solvents are usually used to create the mobile phase for an HPLC experiment, water and an organic liquid (Khopkar, 2009; Meyer, 2010; Xu, 2013). If the substances in the analyte are predicted to be stable at low pH, 0.1% formic acid or trifluoro-acetic acid (TFA) may be included in the mobile phase. The acid is added because many organic molecules are more stable when fully protonated. All organic liquids are less polar than water. A common protocol for variation of the mobile phase during chromatography involves a transition from a mixture that is predominantly water to one that is predominantly organic. In this gradient of mixed solvents, acetonitrile or methanol is often used as the organic component.

A common type of stationary phase for HPLC of organic metabolites consists of unbranched alkanes that contain 18 carbons each. This is known as a *column of C_{18}*. The alkanes are linked to spherical particles. For HPLC the diameter of these particles is between 2.5 and 10.0 μm, whereas for UHPLC it is between 1.7 and 1.8 μm. The interaction of the molecules in the analyte with the stationary phase may be modulated by variation in the composition of the mobile phase as described earlier. If the nonpolar portions of the molecules in the mobile phase associate more tightly with the stationary phase than does a particular type of molecule in the analyte, the rate of migration of this component of the analyte will increase. This increase in rate will decrease the time of retention. An increase in the concentration of organic solvent to 90.0% or more of the mobile phase near the end of the chromatographic protocol "cleans" the column. This method prepares it for application of another analyte.

The typical pressure inside the column during HPLC or UHPLC is between 5000.0 and 15,000.0 kPa, corresponding to between 725.0 and 2174.0 pound-force per square inch (psi). The rate of flow of the mobile phase in HPLC is typically between 1.0 and 2.5 mL min^{-1}, whereas in UHPLC it is between 0.1 and 0.2 mL min^{-1}. The time required for each HPLC assay is between 20.0 and 40.0 min whereas for UHPLC it is between 3.0 and 10.0 min.

Detection and quantification of organic metabolites in the analyte

The most common technique for detection of organic molecules as they exit from the column during HPLC is absorption of UV or visible radiation. Most organic molecules do not absorb visible radiation; however, many organometallic compounds do absorb in the visible range. In a later chapter in which you are guided in designing your own experiments, the phenomenon of fluorescence is discussed. A detector for fluorescence may be included in a device used for HPLC. If one or more of the organic compounds in the analyte is predicted to be fluorescent, this type of detection is an option. To precisely determine the molar mass of each type of molecule in the analyte, a device for *mass spectrometry* (MS) may be linked to the device for HPLC. This creates the hybrid technique of liquid chromatography−mass spectrometry (LC−MS) in which each organic compound in the analyte is examined with MS immediately after it exits the HPLC column. LC−MS is described more extensively in the later chapter that focuses on MS and related techniques.

The data generated by an HPLC experiment is a *chromatogram*. On this type of graph, time is plotted on the *x*-axis and the magnitude of the signal of detection is plotted on the *y*-axis. The position of each peak on the plot represents the time of retention of an organic compound in the analyte and the area under that peak is directly proportional to the quantity of that compound in the analyte. Examples of these type of chromatograms are shown in the next section of this chapter.

5.2   AN EXAMPLE OF THE USE OF HIGH-PERFORMANCE LIQUID CHROMATOGRAPHY

Preparation of samples

The atom of iron that is chelated by a siderophore is in the $+3$ oxidation state and therefore known as iron(III). Chelation of iron(III) by siderophores has been examined in HPLC experiments (D'Onofrio et al., 2010). The siderophore enterobactin is secreted by the bacterium *Escherichia coli* (Pollack & Neilands, 1970; Raymond & Carrano, 1979; Raymond, Dertz, & Kim, 2003). Before it chelates iron(III), enterobactin has a molecular formula of $C_{30}H_{27}N_3O_{15}$ and a molar mass of 669.55 g mol^{-1}. To demonstrate the chelation activity of enterobactin, the following experiment was performed.

A dried sample of enterobactin was obtained from a commercial supplier (Sigma-Aldrich_Co_enterobactin, 2017) and dissolved in dimethyl sulfoxide at a concentration of 1.0 mg mL^{-1}, corresponding to 1.5 mM. An aqueous solution of iron(III) chloride, $FeCl_3$ (aq), was prepared at a concentration of 1.0 mg mL^{-1}, generating ions of iron(III) at 6.0 mM.

The following three aqueous solutions, each with a final volume was 200.0 μL, were prepared:

- Enterobactin (2.0 µg) from the stock solution diluted into 0.1% TFA;
- $FeCl_3$ (aq) (1.6 µg of the compound) from the stock solution diluted into 0.1% TFA;
- Enterobactin (2.0 µg) and $FeCl_3$ (aq) (1.6 µg of the compound) diluted into 0.1% TFA.

For each assay, 10.0 µL of one of the three solutions described above was injected into the device for UHPLC. Thus the three assays included the following amounts of analytes:

- Enterobactin: 100.0 ng, corresponding to 150.0 pmol;
- $FeCl_3$ (aq): 80.0 ng of the compound, corresponding to 480.0 pmol of iron(III);
- Enterobactin (100.0 ng) and $FeCl_3$ (aq) (80.0 ng of the compound). The molar ratio of enterobactin:iron(III) in this sample was 1.0:3.2.

Chromatographic parameters

Note: As was explained in the previous section in this chapter, the most modern version of this technique is typically known as UHPLC. Waters manufactures devices for this technique and owns the trademark for a slightly different abbreviation, UPLC.

The three samples described above were examined with a UPLC device from Waters. The following parameters were used:

Column holding the stationary phase for the chromatography

Stationary phase: particles coated with unbranched alkanes of eighteen carbons (C_{18}).
Internal diameter: 2.1 mm.
Length: 50.0 mm.
Internal volume: 173.0 µL.

Mobile phase for the chromatography

In most cases a mixture of the two solutions described below was used as the mobile phase for chromatography. As will be explained in later chapters, for some analytes the acid is not included.
Solution A: aqueous 0.10% TFA, pH = 1.8.
Solution B: acetonitrile with 0.18% TFA, pH = 2.0.

- Notes regarding characteristics of the solvents.

Although pure water is expected to have a pH of 7.0, water obtained from the typical filtration systems used in laboratories has a pH of ~ 5.0. Pure acetonitrile has a pH of 7.0. Acetonitrile is much less polar than water. The concentration of acetonitrile in the mobile phase is typically increased to 95.0% at the end of the run to displace any nonpolar substances from the stationary phase, preparing it for application of the next sample of analyte.

Rate of flow, pressure, and the composition of the mobile phase during chromatography

A number of different protocols for the timing of the changes in composition of the mobile phase during the chromatography were used. Below are two such protocols.

- Protocol for a 3-min chromatographic assay.
  The rate of flow of mobile phase was 0.2 mL min^{-1}.
  The range of pressure in the column was 8280.0−14,490.0 kPa (1200.0−2100.0 psi).

  0.0 to 2.0 min: 100.0% solution A.
  2.0 to 3.0 min: gradient from 100.0% solution A to 5.0% solution A, 95.0% solution B.

  With the 3-min protocol, analytes such as the iron-chelating siderophores exited the column within the first 2 min, before acetonitrile was introduced.

- Protocol for a 10-min chromatographic assay.

  The rate of flow of the mobile phase was 0.1 or 0.2 mL min^{-1}.
  If the rate was 0.1 mL min^{-1}, the range of pressure in the column was 6210.0−9660.0 kPa (900.0−1400.0 psi).
  If the rate was 0.2 mL min^{-1}, the range of pressure in the column was 8280.0−14,490.0 kPa (1200.0−2100.0 psi).

  0.0 to 10.0 min: gradient from 100.0% solution A to 5.0% solution A, 95.0% solution B

  With the 10-min protocol, analytes such as the *Vibrio* autoinducer 1 and iron-chelating siderophores exited the column within the first 2 min, before the concentration of acetonitrile reached 20.0%.

Detection of analytes by absorption of ultraviolet or visible radiation

The three samples described earlier in this section were evaluated in several distinct UHPLC assays, using the various

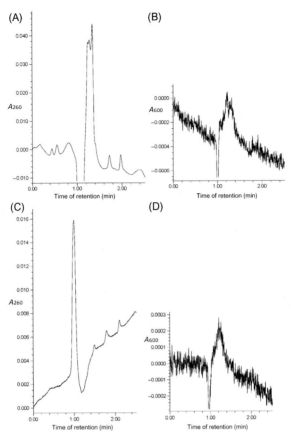

FIGURE 5.1   The chelation of iron(III) by enterobactin, detected with UHPLC and the absorption of ultraviolet and visible radiation. Each panel is a chromatogram of the absorption of analytes as they exited the column. Samples that included enterobactin contained 150.0 pmol of this siderophore and those that included iron(III) had 480.0 pmol of these metal ions. The A_{260} (ultraviolet radiation) is displayed in panels A, C, and E, whereas the A_{600} (visible radiation) is displayed in panels B, D, and F. (A) Enterobactin gave a broad peak of A_{260} that was retained on the column for 1.3 min. (C) Unchelated iron(III) gave a sharp peak of A_{260} was retained for 1.0 min. (E) A mixture of enterobactin and iron (III) gave a sharp peak of A_{260} that was retained for 1.3 min. The lack of a peak with a retention time of 1.0 min in panel E suggests that iron(III) was comigrating with enterobactin, presumably because it had been chelated. (B) Enterobactin gave only a very small peak of A_{600}. (D) Iron(III) gave only a very small peak of A_{600}. (F) A mixture of enterobactin and iron(III) gave a moderate peak of A_{600} that was retained for 1.2 min, close to the time of retention observed for the A_{260} of the mixture. As is shown in a figure in an earlier chapter, the color of a solution of $FeCl_3$ (aq) changes when the iron(III) is chelated by enterobactin. This observation implies that the peak of A_{600} that was detected only for the mixture of enterobactin and iron(III), but not for either molecule alone, was a consequence of chelation.

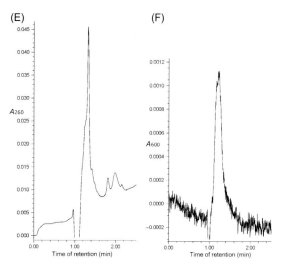

FIGURE 5.1   Continued.

chromatographic parameters described above. The results were consistent in these experiments. The data shown here are from an assay in which the 10-min protocol was used and the rate of flow was $0.1\,\text{mL min}^{-1}$.

The chromatograms recorded for the assay with UV radiation of 260.0 nm show that the peak typically observed for unchelated iron(III) is no longer observed after the addition of enterobactin (Fig. 5.1). Those recorded for the assay with visible radiation of 600.0 nm reveal the appearance of a peak for the mixture that is not observed for either molecule by itself. These two observations suggest that enterobactin-chelated iron(III).

BIBLIOGRAPHY

D'Onofrio, A., Crawford, J. M., Stewart, E. J., Witt, K., Gavrish, E., Epstein, S., … Lewis, K. (2010). Siderophores from neighboring organisms promote the growth of uncultured bacteria. *Chemistry and Biology*, *17*, 254–264. Retrieved from <https://www.ncbi.nlm.nih.gov/pubmed/20338517>.

Khopkar, S. M. (2009). *Basic concepts of analytical chemistry* (3rd ed.). Tunbridge Wells: New Age Science. Retrieved from <http://explore.bl.uk/BLVU1:LSCOP-ALL:BLL01014923277>.

Meyer, V. (2010). *Practical high-performance liquid chromatography* (5th ed.). Chichester, UK: Wiley. Retrieved from <https://lccn.loc.gov/2009052143>.

Pollack, J. R., & Neilands, J. B. (1970). Enterobactin, an iron transport compound from *Salmonella typhimurium*. *Biochemical and Biophysical Research Communications*, *38*, 989–992. Retrieved from <http://www.sciencedirect.com/science/article/pii/0006291X70908193>.

Raymond, K. N., & Carrano, C. J. (1979). Coordination chemistry and microbial iron transport. *Accounts of Chemical Research*, *12*, 183–190. Available from https://doi.org/10.1021/ar50137a004.

Raymond, K. N., Dertz, E. A., & Kim, S. S. (2003). *Enterobactin: An archetype for microbial iron transport*, . *Proceedings of the National Academy of Sciences* (100, pp. 3584–3588). Retrieved from<http://www.pnas.org/content/100/7/3584.abstract>.

Sigma-Aldrich_Co_enterobactin. (2017). *Enterobactin from* Escherichia coli, *item E3910*. St Louis, MO. Retrieved from <http://www.sigmaaldrich.com/catalog/product/sigma/e3910?lang = en&region = US. >.

Xu, Q. A. (2013). *Ultra-high performance liquid chromatography and its applications*. Hoboken, NJ: John Wiley & Sons Inc. Retrieved from <https://lccn.loc.gov/2012035740>.

Chapter 6

Mass spectrometry

6.1 THEORY AND PRACTICE OF MASS SPECTROMETRY

Fundamentals

Although mass spectrometry (MS) is often used to characterize inorganic molecules, this discussion pertains to the study of organic metabolites. The device in which this technique is performed is known as a *mass spectrometer.* It consists of an ionization chamber in which the molecules being examined (the *analyte*) enter the gas phase and acquire electrical charge. These molecular ions are then propelled through the *mass analyzer* component of the device by means of an electrical field (Crews, Rodríguez, & Jaspars, 2010; Khopkar, 2009; Niessen, 2006).

The data that is generated is a *mass spectrum.* The symbol m represents the molar mass of a molecule of analyte in grams per mole and the symbol z represents the number of charges on this molecule. The spectrum is a histogram of the relative amount of each type of analyte that has a particular *mass-to-charge* ratio. The symbol for this ratio is *m/z.* If the value of z is known, then the molar mass of a molecule may be calculated from the measured *m/z.* The molar mass may be calculated to within 0.01 g mol^{-1} and this can provide strong evidence that a particular molecule corresponds to a known structure.

There are several methods available for ionization of molecules of analyte in the various models of mass spectrometers. The *electrospray ionization* method is most commonly used for study of organic metabolites. There are also several types of mass analyzers. Three of the most commonly used analyzers are the *magnetic deflection*, *quadrupole*, and *time-of-flight* (TOF) devices. Only the quadrupole and TOF analyzers are discussed here.

Note: The literary references cited in this chapter include diagrams that will clarify the concepts discussed here.

Purification and Characterization of Secondary Metabolites.
DOI: https://doi.org/10.1016/B978-0-12-813942-4.00006-1

Electrospray ionization

To describe the ions, a neutral molecule of analyte is designated with the uppercase letter "M." An H^+ may be added to, or lost from, each molecule of analyte. This generates a *protonated molecule*, indicated by the notation $+H$, or a *deprotonated molecule* indicated by the notation $-H$. The net charge on the ionic form is denoted by a plus or minus sign in superscript. The protonated form is thus denoted $[M+H]^+$, whereas $[M-H]^-$ represents the deprotonated form [Eqs. (6.1) and (6.2)]. To detect $[M+H]^+$ the mass spectrum is recorded in *positive-ion mode*, whereas *negative-ion mode* is used to detect $[M-H]^-$.

$$M + H^+ \rightarrow [M + H]^+ \qquad (6.1)$$

$$M \rightarrow [M - H]^- + H^+ \qquad (6.2)$$

Solution containing the analyte is injected into an ionization chamber that is maintained at atmospheric pressure. An electric field of between 2.0 and 5.0 kV and a stream of inert nitrogen gas cause *nebulization* of the solution, creating an aerosol. The electric field also introduces electrical charge to the molecules of analyte in the droplets. As the liquid in the droplets evaporates, the molecular ions are released in the gas phase. These ions are then injected into the mass analyzer.

Quadrupole mass analyzers

The chamber of the analyzer is maintained well below atmospheric pressure by means of a vacuum pump. This minimizes interference from atmospheric gases and water vapor. An electric field is created by a cluster of four charged rods (poles), two positive and two negative. A parallel arrangement of these rods creates a *quadrupole* with a gap extending the entire length of the four rods. The rods are charged with direct current (DC) upon which is superimposed a radio-frequency alternating current (RF-AC). This creates a dynamic *quadrupole field* in the gap.

An electrical potential of a few Volts projects the molecular ions in the analyte through the gap. The intensities of the DC and RF-AC in the quadrupole are gradually varied. As the ions are moving through the gap, they oscillate in a plane perpendicular to the length of the rods. For most values of *m/z* in the molecules of analyte, the oscillations will be large enough to cause a collision with one of the rods or exit from the quadrupole, preventing these ions from reaching the detector. For each value of m/z, however, a particular combination of DC and RF-AC will not induce the perpendicular oscillations and the ion will reach the detector. The quadrupole acts as a dynamic filter, restricting access to the detector to one value of m/z at a time. This *scanning* method allows the detector to quantify the amount of molecular ions with each value of m/z.

Time-of-flight mass analyzers

In this variation the chamber through which the ions are pro-pelled is known as the flight tube. As explained above for the chamber of the quadrupole analyzer, the flight tube is main-tained well below atmospheric pressure by means of a vacuum pump. There is no electrical field inside of the tube. The mole-cular ions in an analyte are accelerated in batches by brief (0.25 μs) pulses of an electrical field that is generated only at the entry port of the tube. In each batch, each ion initially has the same amount of kinetic energy. Those with a greater value of m/z take more time to reach the detector than those with a lower m/z. The TOF to reach the detector ranges from 50.0 to 100.0 ms. The TOF increases as m/z increases, therefore an m/z may be calculated for each type of ion from its TOF.

Mass spectrometry linked to chromatography

For a pure substance, good data may be obtained by injecting it directly into an MS device. If the analyte is expected to be a mix-ture, or if it is possible that the purification method did not remove all of the undesired substances, a hybrid technique in which chromatography precedes MS should be used (Crews et al., 2010; Khopkar, 2009; Niessen, 2006). If the chro-matographic parameters are set properly, the substance of inter-est will be separated from other substances. In this way a mass spectrum of the pure substance of interest may be obtained.

A chromatographic device may be physically linked to a mass spectrometer. Thermostable molecules such as fatty acids may be vaporized and chromatographically separated in the gas phase. This is gas chromatography−mass spectrometry. Thermolabile compounds require liquid chromatography−mass spectrometry (LC−MS). In this case high-performance liquid chromatography, or ultra-high performance liquid chromatography, is used. The molecules of analyte are dissolved in a solvent (the *mobile phase*) and this solution migrates through a nonpolar solid matrix (the *stationary phase*). The mobile phase is typically a mixture of water and an organic solvent such as acetonitrile or methanol. The relative concentrations of water and the organic solvent are varied during chromatography. If the analyte is predicted to be stable at low pH, 0.1% formic acid or trifluoro-acetic acid is included in the mobile phase.

In addition to the mass spectrum, a *chromatogram* is generated in an LC–MS experiment. A chromatogram depicts the amount of analyte eluting from the column at a range of times. The amount of time from when a particular type of molecule in the analyte enters the column until it exits is the *retention time*. If the substances in the analyte efficiently absorb UV or visible radiation, spectrophotometry may be used to generate the chromatogram. If absorption of radiation is not an option, *mass detection* is used. If there is a single peak on the chromatogram, a mass spectrum of the substance in this peak is examined. If there are multiple peaks, the mass spectrum of the peak that is believed to be the substance of interest is examined.

6.2 EXAMPLES OF THE USE OF LIQUID CHROMATOGRAPHY–MASS SPECTROMETRY

Preparation of samples of metabolites

Ferrioxamine E

- This substance is a siderophore secreted by *Streptomyces antibioticus* (Cambridge_Crystallographic_Data_Centre, 2017; Van der Helm & Poling, 1976). It is an organometallic compound that includes iron(III). The molecular formula is $C_{27}H_{25}N_6O_9Fe$ and its molar mass is 653.53 g mol^{-1}.
- A sample of 3.0 mg was obtained from a commercial supplier (Sigma-Aldrich_Co., 2017) and then dissolved in 20.0% methanol at a concentration of 10.0 mg mL^{-1}.
- Appropriate dilutions were made into 20.0% methanol and then 5.0 μL that contained 153.0 pmol (100.0 ng) was injected into the LC–MS device.

Tagetitoxin

- This substance is a phytotoxin secreted by *Pseudomonas syringae* pathovar *tagetis*. It contains a primary amine and a phosphate (Aliev, Karu, Mitchell, & Porter, 2016; Mitchell, Coddington, & Young, 1989; Mortimer, Aliev, Tocher, & Porter, 2008). At a pH of 5.0, the amine will be protonated, the phosphate will be ionized, and the molar mass will be 416.294 g mol^{-1}. The molecular formula is $C_{11}H_{17}N_2O_{11}PS$.

- An aqueous solution of tagetitoxin at 600.0 μM (250.0 μg mL^{-1}) was obtained from a commercial supplier (Epicentre_An-Illumina-Company, 2017).
- An aliquot of the stock was diluted 10-fold in deionized water and 5.0 μL that contained 300.0 pmol (125.0 ng) was injected into the LC−MS device.

Chromatographic parameters

Ferrioxamine E

- A mixture of two solutions was used as the mobile phase for chromatography.

    solution A: aqueous 0.1% formic acid
    solution B: acetonitrile with 0.18% formic acid

    Acetonitrile is much less polar than water and it has two functions in this protocol. A gradual increase in the percentage of solution B in the mobile phase that flows through the column ensures that the analyte will elute within the allotted time (10.0 min in this case). The concentration of acetonitrile is raised to 95.0% at the end of the run because it displaces any nonpolar substances from the solid phase in the column, preparing it for application of the next sample.

    Formic acid or trifluoro-acetic acid is included in the mobile phase for chromatography of many types of analytes. This is probably because these analytes are more stable when fully protonated.

- The rate of flow of the mobile phase was 0.5 mL min^{-1}. The composition was as follows.

    From 0.0 to 7.0 min, a gradient from 90.0% solution A, 10.0% solution B to 5.0% solution A, 95.0% solution B.

    From 7.0 to 10.0 min it stayed constant at 5.0% solution A, 95.0% solution B.

Note regarding the pH of the mobile phases: Although pure water is expected to have a pH of 7.0, water obtained from the typical filtration systems used in laboratories has a pH of ∼5.0. Pure acetonitrile has a pH of 7.0.

Tagetitoxin

- Tagetitoxin is unstable at low pH (Mitchell & Durbin, 1981). The gradient of solutions in the mobile phase during chromatography was the same as that used for ferrioxamine E except that formic acid was not included. In this case the pH of the mobile phase was 5.0.

Interpretation of chromatograms and mass spectra

The data collected for ferrioxamine E (Fig. 6.1) and tagetitoxin (Fig. 6.2) are presented here. The interpretations of the results are in the legends below the figures.

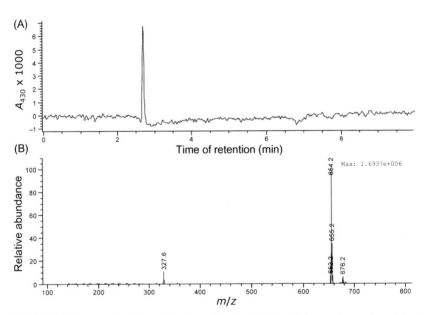

FIGURE 6.1 Examination of ferrioxamine E with liquid chromatography–mass spectrometry in positive ion mode. (A) A chromatogram showing the A_{430} of analyte molecules exiting the column as a function of time. Iron(III) bound to ligand is known to absorb radiation with a wavelength of 430.0 nm. The peak at 2.7 min includes ferrioxamine E. (B) The molecules of analyte exiting the column at 2.7 min were evaluated with mass spectrometry. This histogram shows the relative abundance of analyte molecules in this peak with different values of m/z. Ferrioxamine E, denoted by the symbol "M" is detected in three forms. The largest bar, at 654.2, is the protonated form $[M + H]^+$. The bar at 327.6 is the doubly protonated form $[M + 2H]^{2+}$ and the bar at 676.2 is an adduct with an ion of sodium $[M + Na]^+$. Because the molar mass of ferrioxamine E has been reported as 653.53 g mol^{-1}, these values of m/z are as expected.

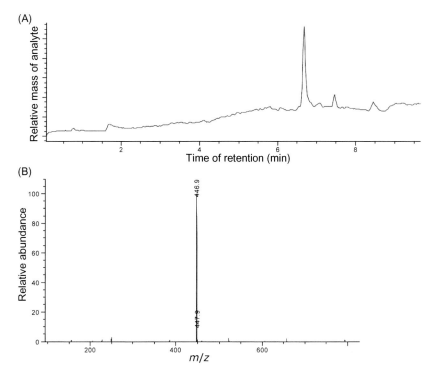

FIGURE 6.2 Examination of tagetitoxin with liquid chromatography–mass spectrometry in negative ion mode. (A) A chromatogram showing the mass of analyte exiting the column as a function of time. The peak at 6.7 min includes tagetitoxin. (B) The molecules of analyte exiting the column at 6.7 min were evaluated with mass spectrometry. This histogram shows the relative abundance of analyte molecules in this peak with different values of m/z. The bar at 446.9 is presumably the deprotonated $[M-H]^-$ form of tagetitoxin. This value is ∼8.0% greater than the expected value but close enough to make this conclusion reasonable.

BIBLIOGRAPHY

Aliev, A. E., Karu, K., Mitchell, R. E., & Porter, M. J. (2016). The structure of tagetitoxin. *Organic and Biomolecular Chemistry*, *14*, 238–245. Retrieved from < https://www.ncbi.nlm.nih.gov/pubmed/26517805 >.

Cambridge_Crystallographic_Data_Centre. (2017). *FEROXE10*. Reference code FEROXE10. Cambridge Structural Database. Cambridge, England. Retrieved from < https://www.ccdc.cam.ac.uk/structures-beta/Search?Ccdcid = feroxe10 >.

Crews, P., Rodríguez, J., & Jaspars, M. (2010). *Organic structure analysis* (2nd ed.). New York, NY: Oxford University Press. Retrieved from < https://lccn.loc.gov/2009018383 >.

Epicentre_An-Illumina-Company. (2017). *Tagetin^{TM} RNA polymerase inhibitor*. Madison, WI. Retrieved from < http://www.epibio.com/docs/default-source/protocols/tagetin-rna-polymerase-inhibitor.pdf?sfvrsn = 12 >.

Khopkar, S. M. (2009). *Basic concepts of analytical chemistry* (3rd ed.). Tunbridge Wells: New Age Science. Retrieved from <http://explore.bl.uk/BLVU1:LSCOP-ALL:BLL01014923277>.

Mitchell, R. E., Coddington, J. M., & Young, H. (1989). A revised structure for tagetitoxin. *Tetrahedron Letters, 30*, 501−504. Retrieved from <http://www.sciencedirect.com/science/article/pii/S0040403900952390>.

Mitchell, R. E., & Durbin, R. D. (1981). Tagetitoxin, a toxin produced by *Pseudomonas syringae* pv. tagetis: Purification and partial characterization. *Physiological Plant Pathology, 18*, 157−168. <http://www.sciencedirect.com/science/article/pii/S0048405981800379>.

Mortimer, A. J., Aliev, A. E., Tocher, D. A., & Porter, M. J. (2008). Synthesis of the tagetitoxin core via photo-Stevens rearrangement. *Organic Letters, 10*, 5477−5480. Retrieved from <https://www.ncbi.nlm.nih.gov/pubmed/18973329>.

Niessen, W. M. A. (2006). *Liquid chromatography−mass spectrometry* (3rd ed.). Boca Raton, FL: CRC/Taylor & Francis. Retrieved from <https://lccn.loc.gov/2006013709>.

Sigma-Aldrich_Co. (2017). *Ferrioxamine E, product 38266, specifications*. St Louis, MO. Retrieved from <https://www.sigmaaldrich.com/catalog/DataSheetPage.do?brandKey=SIAL&symbol=38266>.

Van der Helm, D., & Poling, M. (1976). The Crystal structure of ferrioxamine E. *Journal of the American Chemical Society, 98*, 82−86. Available from https://doi.org/10.1021/ja00417a014.

Chapter 7

Nuclear magnetic resonance spectroscopy

7.1 THEORY AND PRACTICE OF NUCLEAR MAGNETIC RESONANCE SPECTROSCOPY

The importance of nuclear magnetic resonance for structural studies

Although chromatography, mass spectrometry, and quantification of the absorption of radiation provide extensive knowledge of molecular characteristics, more sophisticated technology is necessary to completely solve a structure. Nuclear magnetic resonance (NMR) spectroscopy and X-ray crystallography are the most commonly used methods for solving the structures of metabolites. NMR assays elucidate the bonding arrangement of atoms in a molecule; however, crystallography provides additional information. For example, the spatial configuration around a chiral center, the lengths of bonds and the angles between adjacent bonds are provided by crystallography.

Crystallography, however, has a number of limitations. The most significant is the need for single crystals. Although the known physical properties of a molecule, such as solubility, are used to design a scheme for crystallization, some compounds are very difficult to crystallize (Glazer, 2016; Massa, 2004; Stout & Jensen, 1989). If many attempts have to be made to get crystals formed, tens or hundreds of milligrams of the compound may be required. Sufficient NMR spectroscopic data for metabolites can often be obtained from a solution containing just a few milligrams.

Note: The literary references that are cited in the following sections have diagrams that clarify the concepts discussed here.

Purification and Characterization of Secondary Metabolites.
DOI: https://doi.org/10.1016/B978-0-12-813942-4.00007-3

The magnetic properties of nuclei and molecules

A charged particle in motion creates a magnetic field. Quantum mechanics theory of atomic structure states that each electron in an atom may have one of two types of *spin*, and we are told to visualize this as the two possible directions of rotation of a sphere around a single axis. The same concept may be applied to the atomic nucleus. Although neutrons are uncharged, the protons of the nucleus have positive charge. Thus for some nuclei, spin creates a *local magnetic field* (Crews, Rodríguez, & Jaspars, 2010; Keeler, 2010; Richards & Hollerton, 2011; Rule & Hitchens, 2006). When the direction of spin is considered, this field can be in either of two orientations. Thus the local field can be described as either of two vectors or *magnetic moments*. In the absence of an external magnetic field, the potential energy of the nucleus does not depend on the direction of spin. The change that occurs when a field is applied is described below.

Not all nuclei exhibit a magnetic moment. In quantum mechanics theory, the local magnetic field of nuclei is said to arise from *spin angular momentum*, described by the *spin quantum number* (I). Nuclei with an odd mass number such as ^1H, ^{13}C, and ^{15}N have this property. Nuclei with an odd number of protons, such as ^2H (*deuterium*, designated D in formulas) and ^{14}N, also have this property. The predominant carbon and oxygen isotopes, ^{12}C and ^{16}O, do not have this feature.

NMR spectroscopy detects only those nuclei with spin angular momentum. In *proton NMR*, ^1H (with $I = \frac{1}{2}$) and ^2H (with $I = 1$) are both detected. However, because ^1H is the predominant form of hydrogen in nature, it is this isotope that is detected in the molecule being examined (the *analyte*). Solvents used to dissolve analytes for NMR are *deuterated* (atoms of ^1H have been replaced with ^2H) to prevent the solvent from generating strong peaks in the spectrum. In some cases, as will be discussed later, a *residual ^1H solvent* signal is present in the spectrum. Although ^2H peaks are not seen in the spectrum, the magnetic property of this isotope is used as a reference during an NMR spectroscopic experiment. This will also be discussed later.

Effect of continuous application of an external magnetic field on ^1H nuclei

The primary electromagnet of a typical NMR spectrometer consists of a coil of superconductor material that is maintained in liquid helium at a temperature of 4.0 K. It creates a static magnetic field of 9.4 or 11.75 T. This corresponds to 9.4 or 11.75×10^4 G. This magnetic field is designated B_0 (Crews et al., 2010; Keeler, 2010; Richards & Hollerton, 2011; Rule & Hitchens, 2006). The solution of *analyte* (substance to be evaluated) is placed in a tube in the center of the *probe* assembly that is in the midst of the B_0. The probe includes the *excitation coil* that is described below. Insulation keeps the probe at ambient temperature so that the analyte remains in the liquid state.

When a B_0 field is applied to a dissolved analyte, the spin axis of each ^{1}H or ^{2}H nucleus takes on one of two orientations, such that the magnetic moment is in the direction of B_0 (α nuclei) or directly opposed to B_0 (β nuclei). The α nuclei have lower potential energy (*ground state*) than the β nuclei (*excited state*) and the difference between the two energy levels is ΔE. If the only external field present is the B_0, the nuclei in the analyte molecules will assume an *equilibrium* distribution. There will be slightly more α nuclei than β nuclei (~ 1 in 10^6). As will be explained below, the transverse pulses from another coil may be used to boost a small percentage of the ground state nuclei to the excited state. After the pulse, the same percentage of nuclei gradually return to ground state and the analyte is back at equilibrium.

The Greek letter nu (ν) represents a frequency. Frequency is quantified in reciprocal seconds (Hz). Planck's constant is equal to 6.626×10^{-34} J s and is denoted by the symbol h. Eq. (7.1) depicts the relationship between the energy and the frequency of radiation emitted by nuclei returning to ground state from the excited state.

$$\Delta E = h\,\nu \qquad (7.1)$$

The *Larmor equation* (Eq. (7.2)) shows that the *resonance frequency* of a particular isotope is directly proportional to the B_0. This version of the equation is for nuclei of ^{1}H, thus the term ν (^{1}H) represents the resonance frequency. The Greek letter gamma (γ) represents the *gyromagnetic ratio*. The value of γ is different for each isotope. For ^{1}H it is 2.6753×10^8 rad s^{-1} T^{-1} whereas for ^{2}H it is a much smaller value (Rule & Hitchens, 2006).

$$\nu(^1\text{H}) = \frac{\gamma B_0}{2\pi} \qquad (7.2)$$

During the course of an NMR experiment, it is ideal to keep the B_0 constant and homogeneous throughout the chamber holding the sample. Since the field emanating from any electromagnet tends to drift a bit, some additional magnets, not nearly as powerful as that described above are used to compensate for any inhomogeneity in B_0. These smaller magnets are called *shims* and the process of adjusting their intensity is *shimming*. The deuterated solvents typically used to dissolve NMR analytes are well characterized. As mentioned above the value of the gyromagnetic ratio is lower for ^2H than for ^1H. The means that the ν (^2H) is well outside the range of values that are recorded in a typical ^1H NMR spectrum of an organic molecule. Thus, without any disturbance of the ^1H spectrum, the spectrometer may *lock* on the ^2H signal and use a feedback system to adjust shim intensity and keep the B_0 homogeneous.

Effect of pulsed application of an external magnetic field on ^1H nuclei

Pulses of *radio frequency* (RF) electromagnetic radiation are emitted from the *excitation coil* in the *transmitter* in the spectrometer. The pulses are brief (up to 20.0 μs). Each pulse is not a single frequency, rather a spectrum of frequencies. Each frequency in each pulse of RF creates a magnetic field known as B_1 that is perpendicular to B_0. B_1 is a *transitory field* and is also known as the *excitation field*. B_1 can induce a transition in the magnetic spin of the nucleus of an atom of ^1H, changing its energy from the ground state to the excited state. After the pulse has ended, the excitation coil becomes a *receiver* that quantifies the radiation emitted by the nuclei returning to ground state. The receiver therefore detects the ν (^1H) of the various hydrogen atoms. This is the *acquisition* step and the mathematical process by which the various values of ν (^1H) are calculated is described below. The process is repeated and data compiled to improve the signal-to-noise ratio. Each repeat is called a *scan* or a *transient*.

For example, in a device generating a B_0 of 11.75 T, the ν (^1H) is approximately 500.0 MHz. An *NMR spectrum* shows how the ν (^1H) varies slightly amongst the different hydrogen atoms in an organic molecule. The spectrum shows the intensity of resonance for each ^1H at a particular *chemical shift*.

Because the variation is so slight, the *x*-axis of the graph that displays the signals is calibrated in parts per million (ppm). An NMR spectrometer is usually described by the frequency of resonance for a particular isotope rather than the strength of the B_0. The device mentioned above would be known as a 500.0 MHz ^{1}H NMR spectrometer.

Acquisition of the free induction decay as ^{1}H nuclei return to equilibrium

The receiver monitors the *free induction decay* (FID) that occurs over a period of between 50.0 ms and 3.0 s. The FID is energy released from the small percentage of [1]*H* nuclei returning from the excited state to the ground state. It is detected as a set of sinusoidal waves that gradually dissipate to baseline. This return to ground state is also known as *relaxation* of the nuclei. Each wave has a distinct frequency that corresponds to the ν (^{1}H) of one or more hydrogen atoms in the analyte. The various waves start simultaneously, so the data are observed as many overlapping waves that is known as a *time domain* plot. This is converted to a *frequency domain* plot by a mathematical operation known as the *Fourier transform*. In a frequency domain plot, each peak represents a distinct frequency. This is the NMR spectrum.

Effect of local magnetic fields on relaxation: interference from paramagnetic substances

The rate of relaxation of a population of nuclei is increased by the *local magnetic fields* generated by the nuclei and electrons of the analyte and any contaminants (Crews et al., 2010; Field, Sternhell, & Kalman, 2013; Keeler, 2010). Any moving, charged particle has a magnetic moment. Thus both electrons and protons have a magnetic moment but the moment of an electron is much greater than that of a proton. Like protons, electrons can have two types of spin. If any of the atoms in the analyte or any contaminating molecules have more electrons with one type of spin than the other, the *unpaired electrons* will project a local magnetic field (Atkins, Jones, & Laverman, 2016). Such a substance is described as *paramagnetic*. Both iron(III) that is associated with some organic molecules, and molecular oxygen (O_2) that may enter a solution of analyte from the atmosphere, are paramagnetic. Acceleration of the relaxation of nuclei by these substances can cause widening of the peaks in the spectrum and a decrease in resolution. For this reason it is best to assay analytes that have no iron(III) or O_2.

Calculation of the chemical shifts for nuclei of ^1H and generation of a resonance spectrum

Three aspects of the method for analysis of ^1H NMR data assure scientists that the spectra obtained for a particular compound in a particular solvent will be nearly identical regardless of the design of the spectrometer.

- Values of ν (^1H) in the analyte are measured relative to ν (^1H) in a reference compound, usually tetramethylsilane (TMS). The 12 hydrogen atoms in TMS are equivalent and the reference frequency is denoted ν $(^1H)_{ref}$ (Keeler, 2010; Richards & Hollerton, 2011; Rule & Hitchens, 2006).
- The difference between a particular ν (^1H) in the analyte and the ν $(^1H)_{ref}$ is expressed as a fraction of the ν $(^1H)_{ref}$.
- The ν (^1H) of the residual ^1H solvent is used to reference the spectrum to the ν $(^1H)_{ref}$ at 0.0 ppm, according to the method recommended by the International Union of Pure and Applied Chemistry (IUPAC) and the protocols of the manufacturer of the spectrometer (Crews et al., 2010; Field et al., 2013; Richards & Hollerton, 2011). Residual ^1H solvent arises from exchange of a small percentage of the ^2H in the solvent with ^1H in the analyte.

The method described above compensates for the variation in strength of the B_0 and other features in different spectrophotometers. For each ^1H nucleus in the analyte, this difference in the frequency of resonance is its *chemical shift*. A chemical shift is designated with a lowercase Greek letter delta, δ. Because a chemical shift represents only a very small percentage of the ν $(^1H)_{ref}$, it is expressed in parts per million that is abbreviated as ppm (Eq. (7.3)).

$$\delta(^1H) \, (ppm) = \frac{\nu(^1H) - \nu(^1H)ref}{\nu(^1H)ref} \times 10^6 \qquad (7.3)$$

NMR spectra are presented as peaks of chemical shifts. The graphical plot has the zero point on the right end of the x-axis, thus greater shifts place peaks farther to the left. The area of each peak is proportional to the number of ^{1}H nuclei in the analyte that has that particular shift.

Effect of molecular structure on chemical shifts for nuclei of ^{1}H

As explained above, electrons generate a local magnetic field. This field partially counteracts the effect of the B_0 field on nuclei. Due to this *shielding* effect, the greater the electron density surrounding the nucleus, the smaller the ΔE between α and β nuclei when B_0 is applied. TMS is used as the reference compound because the 12 methyl hydrogens are more shielded than hydrogens in most other organic compounds. Since each ν (^{1}H) is directly proportional to the amount of B_0 that gets through the shielding, the values of ν (^{1}H) in most compounds will be greater than that of ν (^{1}H)$_{ref}$. Most chemical shifts are therefore positive numbers.

Electron-withdrawing groups near an ^{1}H nucleus will decrease shielding and increase the chemical shift. Nitrogen, oxygen, aromatic rings, carbon—carbon double bonds, and carbon—carbon triple bonds near an ^{1}H nucleus will increase the shift. The references cited above provide examples of chemical shifts in the ^{1}H NMR spectra of many organic compounds.

Effect of spin—spin coupling on an ^{1}H nuclear magnetic resonance spectrum

Spin—spin coupling is due to the interaction of the local magnetic fields of two or more ^{1}H nuclei in an organic molecule (Crews et al., 2010; Field et al., 2013; Keeler, 2010; Richards & Hollerton, 2011; Vollhardt & Schore, 2014). It occurs between nuclei that are within three bonds of one another, thus the most common example is hydrogens on adjacent carbons. This coupling results in *splitting* of a peak in the spectrum to generate a *doublet*, *triplet*, or more complex cluster. If the number of new peaks that arise due to splitting cannot be determined precisely, the term *multiplet* is used. A figure showing a sample spectrum, and a table of chemical shifts deduced from that spectrum, is presented later in this chapter. These data provide an example of the splitting of peaks. The references cited above in this paragraph provide more extensive discussions of this phenomenon and how it is used to solve molecular structures.

7.2   AN EXAMPLE OF THE USE OF NUCLEAR MAGNETIC RESONANCE SPECTROSCOPY

To provide an example of an NMR spectrum and how it is interpreted, desferrioxamine B was examined. Desferrioxamine B is a siderophore that is secreted by the bacterium *Streptomyces pilosus* (Borgias, Hugi, & Raymond, 1989; Dhungana, White, & Crumbliss, 2001) and bacterial species in the genus *Micromonospora* (Simionato et al., 2006). The prefix *des* on the name means that it refers to the siderophore without iron(III) bound. After chelation of iron(III) it is described as

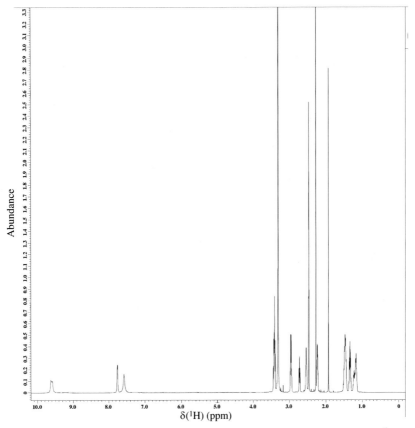

FIGURE 7.1   The nuclear magnetic resonance spectrum for atoms of ^1H in desferrioxamine B mesylate. Chemical shifts are detected as the displacement of peaks (in ppm) from the resonance of the ^1H in the methyl groups of tetramethylsilane at 0.0 ppm. This reference point is near the right end of the x-axis, therefore shifts increase to the left. Assignment of the shifts to the functional groups in desferrioxamine B is summarized in Table 7.1.

ferrioxamine B. The *des* form was examined in this case because iron(III) is paramagnetic and its presence in the analyte will cause a decrease in resolution in the NMR spectrum.

An aliquot of solid desferrioxamine B mesylate was purchased from a commercial supplier (Sigma-Aldrich_Co_ desferrioxamine_B, 2017). Mesylate is a counterion not naturally associated with this siderophore. The siderophore was dissolved in deutero dimethyl sulfoxide (D_6-DMSO) at a concentration of 6.6 mg mL^{-1} (10.0 mM) and 0.6 mL that contained 4.0 mg (6.0 μmol) was assayed with NMR spectroscopy. A single-pulse ^1H spectrum was obtained using a model JNM-ECA500 spectrometer from JEOL. This device generated a B_0 (magnetic field) of 11.75 T. Thus the resonance frequency for atoms of ^1H, denoted as ν (^1H), was approximately 500.0 MHz. The slight variations in the frequency of resonance for atoms of ^1H in different parts of desferrioxamine B are the chemical shifts denoted as δ (^1H), quantified in parts per million (ppm) and depicted in the spectrum (Fig. 7.1). Assignments of the values of δ (^1H) to the structure of desferrioxamine B are provided here (Table 7.1 and Fig. 7.2).

TABLE 7.1 Chemical shifts (ppm) for atoms of ^1H in the functional groups of desferrioxamine B mesylate, quantified by nuclear magnetic resonance spectroscopy.

		Source of data[a,b]		
	Number of H atoms	T.E. Crowley[c]	Borgias et al. (1989) and D'Onofrio et al. (2010)	Prediction in textbooks for this functional group[d]
⸺CH_2⸺	16	1.17–1.47 quin, sep, m	1.2–1.6	1.3–1.8 n
⸺CH_3	3	1.93 s	2.0	2.1 s
⸺CH_2⸺adjacent to C$=$O	8	2.23 quar	2.3–2.6	2.2–2.7 t
$CH_3SO_3^-$ (mesylate counterion)	3	2.27 s	2.4	2.5 s
Dimethyl sulfoxide (residual ^1H solvent)	1–6	2.46 quin		2.5 quin
⸺CH_2⸺ adjacent to N in amide	4	2.54–3.42 t, quar, m	2.8–3.5	2.7–3.5 sex
				(Continued)

TABLE 7.1 (Continued)

	Source of data[a,b]			
	Number of H atoms	T.E. Crowley[c]	Borgias et al. (1989) and D'Onofrio et al. (2010)	Prediction in textbooks for this functional group[d]
——CH$_2$—— adjacent to N in hydroxamic acid	6	2.54–3.42 t, quar, m	2.8–3.5	2.7–3.5 t
——CH$_2$—— adjacent to primary amine	2	2.54–3.42 t, quar, m	2.8–3.5	2.7–3.5 m
H$_2$O (contaminant)	2	3.32 s		3.5 s
——NH—— in amide	2	7.76 m	7.8–7.9	6.0–9.5 t
——NH$_3^+$	3	7.58 s	7.7	3.3–4.8 t
——NOH——	3	9.61 m	9.6–9.7	1.0–5.3 t

[a]Desferrioxamine B mesylate was dissolved in D$_6$-DMSO. Exchange of ^{1}H from desferrioxamine B mesylate, or contaminating H$_2$O, with ^{2}H in a small percentage of molecules of D6-DMSO generates residual ^{1}H solvent. The resonance of this residual ^{1}H solvent is used to reference the ^{1}H spectrum with the methyl resonance of TMS at 0.0 ppm, according to the method recommended by the IUPAC and the protocols of the manufacturer of the spectrometer (Crews et al., 2010; Field et al., 2013; Richards & Hollerton, 2011).
[b]Abbreviations: s = singlet, t = triplet, quar = quartet, quin = quintet, sex = sextet, sep = septet, n = nonet, m = multiplet (>9 peaks).
[c]Not previously published.
[d]Crews et al. (2010), Field et al. (2013), Richards and Hollerton (2011), and Taber (2007).

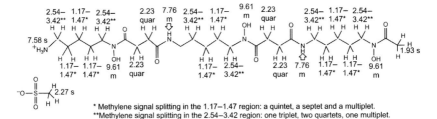

* Methylene signal splitting in the 1.17–1.47 region: a quintet, a septet and a multiplet.
**Methylene signal splitting in the 2.54–3.42 region: one triplet, two quartets, one multiplet.

FIGURE 7.2 Assignment of the chemical shifts for atoms of ^{1}H to the structure of desferrioxamine B. Peak-splitting notation: s = singlet, quar = quartet, m = multiplet.

The data summarized here for atoms of ^{1}H in the functional groups of desferrioxamine B are consistent with previously published spectra for this molecule and predictions in textbooks for these groups in any molecule.

BIBLIOGRAPHY

Atkins, P. W., Jones, L., & Laverman, L. (2016). *Chemical principles: The quest for insight* (7th ed.). New York, NY: W.H. Freeman: Macmillan Learning. Available from https://lccn.loc.gov/2015951706.

Borgias, B., Hugi, A. D., & Raymond, K. N. (1989). Isomerization and solution structures of desferrioxamine B complexes of aluminum(3 +) and gallium (3 +). *Inorganic Chemistry*, *28*, 3538−3545. Available from https://doi.org/ 10.1021/ic00317a029.

Crews, P., Rodríguez, J., & Jaspars, M. (2010). *Organic structure analysis* (2nd ed.). New York, NY: Oxford University Press. Available from https://lccn.loc.gov/ 2009018383.

D'Onofrio, A., Crawford, J. M., Stewart, E. J., Witt, K., Gavrish, E., Epstein, S., ... Lewis, K. (2010). Siderophores from neighboring organisms promote the growth of uncultured bacteria. *Chemistry and Biology*, *17*, 254−264. Available from https://www.ncbi.nlm.nih.gov/pubmed/20338517.

Dhungana, S., White, P. S., & Crumbliss, A. L. (2001). Crystal structure of ferrioxamine B: A comparative analysis and implications for molecular recognition. *Journal of Biological Inorganic Chemistry*, *6*, 810−818. Available from https://www.ncbi.nlm.nih.gov/pubmed/11713688.

Field, L. D., Sternhell, S., & Kalman, J. R. (2013). *Organic structures from spectra* (5th ed.). Chichester, West Sussex: Wiley. Available from https://lccn. loc.gov/2012046033.

Glazer, A. M. (2016). *Crystallography: A very short introduction* (1st ed.). New York, NY: Oxford University Press. Available from https://lccn.loc.gov/ 2015958968.

Keeler, J. (2010). *Understanding NMR spectroscopy* (2nd ed.). Chichester, UK: John Wiley and Sons. Available from https://lccn.loc.gov/2009054393.

Massa, W. (2004). *Crystal structure determination* (2nd completely updated ed.). Berlin; New York, NY: Springer. Available from https://lccn.loc.gov/ 2003069465.

Richards, S. A., & Hollerton, J. C. (2011). *Essential practical NMR for organic chemistry*. Chichester, West Sussex, UK: John Wiley. Available from https:// lccn.loc.gov/2010033319.

Rule, G. S., & Hitchens, T. K. (2006). *Fundamentals of protein NMR spectroscopy*. Dordrecht: Springer. Available from https://lccn.loc.gov/ 2008447860.

Sigma-Aldrich_Co_desferrioxamine_B. (2017). *Desferrioxamine B, sku D9533*. St Louis, MO. http://www.sigmaaldrich.com/catalog/product/sigma/d9533? lang = en&region = US.

Simionato, A., de Souza, G., Rodrigues-Filho, E., Glick, J., Vouros, P., & Carrilho, E. (2006). Tandem mass spectrometry of coprogen and deferoxamine hydroxamic siderophores. *Rapid Communications in Mass Spectrometry*, *20*, 193−199. Available from http://www.ncbi.nlm.nih.gov/entrez/query.fcgi? cmd = Retrieve&db = PubMed&dopt = Citation&list_uids = 16345131.

Stout, G. H., & Jensen, L. H. (1989). *X-ray structure determination: A practical guide* (2nd ed.). New York, NY: Wiley. Available from https://lccn.loc.gov/ 88027931.

Taber, D. F. (2007). *Organic spectroscopic structure determination: A problem-based learning approach*. New York, NY: Oxford University Press. Available from https://lccn.loc.gov/2006035525.

Vollhardt, K. P. C., & Schore, N. E. (2014). *Organic chemistry: Structure and function* (7th ed.). New York, NY: W.H. Freeman and Company. Available from https://lccn.loc.gov/2013948560.

Chapter 8

X-ray crystallography

Note: There are diagrams in all of the literary references cited in this chapter that will clarify the concepts discussed here.

8.1 THE IMPORTANCE OF CRYSTALLOGRAPHY

Although electrophoresis, chromatography, and mass spectrometry provide extensive knowledge of molecular characteristics, elucidation of the precise spatial arrangement of atoms and the lengths of chemical bonds requires more advanced technology. X-ray crystallography and nuclear magnetic resonance (NMR) spectroscopy are the most commonly used methods for solving the structure of an organic or organometallic molecule. Although there are some advantages to using NMR spectroscopy, the spatial arrangement of atoms around a chiral center is more readily determined by the X-ray method than it is by NMR spectroscopy. X-ray crystallography can provide a structure at a resolution of 0.83 Å. (Scientists use the International System of units for most measurements. Although the Ångström is not included in SI, it is commonly used by crystallographers. It is equal to 1.0×10^{-10} m or one-tenth of a nanometer.)

8.2 GROWING CRYSTALS OF AN ORGANIC METABOLITE

A single crystal of a pure substance is an orderly array of molecules. The arrangement of the molecules is the *lattice* of the crystal. For examination of an organic metabolite with X-ray diffraction, single crystals that have a size between 0.25 and 1.0 mm are needed. There are several methods for growing crystals (Glazer, 2016; Laudise, 1970; Massa, 2004; Stout & Jensen, 1989). In most of these methods, the substance is dissolved in a solvent in which it is moderately soluble. This is between 0.1 and 10.0 mg mL^{-1} and this corresponds to between 0.1 and 10.0 mM for a molecule that is 1000.0 g mol^{-1}. If the initial concentration

Purification and Characterization of Secondary Metabolites.
DOI: https://doi.org/10.1016/B978-0-12-813942-4.00008-5
© 2020 Elsevier Inc. All rights reserved.

is not high enough, it is unlikely that crystals of the appropriate size will form. If the substance is too soluble in the initial solvent, it may be difficult to get it to precipitate.

Most crystal-growing strategies involve a gradual change in one or more aspects of the solute and/or solvent. The following are examples of such methods.

- The solution is warmed to a temperature just below the boiling point of the solvent, then allowed to cool very slowly back to ambient temperature.
- The concentration of the solute is gradually increased by allowing the solvent to evaporate at ambient temperature.
- A solvent exchange is performed by *vapor diffusion*.
- A solvent exchange is performed by *layering*.

The "exchange" in the last two methods is from the initial solvent to one in which the substance is less soluble.

An initial characterization of the lattice of a crystal is performed with visible light and a microscope having components that can detect *polarized light*. If *birefringence* of light as it passes through a crystal is observed, this is evidence that it is an *anisotropic* crystal (Laudise, 1970; Massa, 2004; Nikon_Instruments_Inc., 2017; Stout & Jensen, 1989). This phenomenon is also called *double refraction*. Crystals that have any type of lattice other than *cubic* are anisotropic. These properties of crystals and the assay to identify anisotropic crystals are discussed in more detail in the later chapter that includes exercises for characterization of iron-chelating molecules.

8.3 THE THEORY OF X-RAY DIFFRACTION

While refraction of visible light occurs in crystals as described earlier, *diffraction* requires radiation with a wavelength (λ) close to that of the interatomic distances typically found in crystals. This distance is between 1.0 and 3.0 Å. An X-ray that consists of electrons propelled through a vacuum tube and deflected off a molybdenum (Mo) or copper (Cu) anode is usually used to examine a crystal of a metabolite. Although an X-ray is a stream of particles moving through space, it may be described as a wave. The radiation emitted by an anode made of molybdenum has a λ of 0.71 Å, whereas that emitted by an anode made of copper has a λ of 1.54 Å.

Diffraction is observed as reflected X-rays interspersed with regions of weak or no radiation. This phenomenon is due to a combination of *constructive interference* (addition of waves that are *in phase*) and *destructive interference* (canceling of waves that are *out of phase*). The atoms in the organic molecules in a crystal can be thought of as points embedded in a set of *reflection planes* that are parallel to one another. As a crystal is rotated, the angle between the path of the X-rays and this set of planes changes. This is the *angle of incidence* and it is symbolized with the Greek letter theta, θ. The distance between adjacent reflection planes is symbolized as d and it also changes as the crystal is rotated.

The three variables mentioned earlier determine under what circumstances diffraction in the waves of X-ray that are reflected off the planes of atoms will occur. The *order of diffraction* is symbolized as n. *Bragg's law* (Eq. (8.1)) describes the relationship between the three variables and allows for calculation of n.

$$2d\sin\theta = n\lambda \qquad (8.1)$$

If the values of the three variables are such that n in the Bragg equation is a positive integer, constructive interference is occurring. Other values of n indicate destructive interference or a weak signal. The value of $\sin\theta$ increases as the value of θ approaches 90.0°, therefore n increases as θ approaches 90.0°.

8.4 MOUNTING THE CRYSTAL ON AN X-RAY DIFFRACTOMETER AND COLLECTING THE DATA

The crystal is mounted in oil, within a nylon loop, then placed on a *goniometer* that allows for rotation of the crystal around all three axes. Liquid nitrogen keeps the crystal at a temperature of 100.0 K to prevent melting and minimize atomic vibrations. A microscope is included to align the crystal mount with the waves of X-rays. A charge-coupled device (CCD) functions as the detector. The CCD captures the reflected X-rays and diffraction data are recorded by computational software.

8.5 IDENTIFYING THE LATTICE SYSTEM AND BRAVAIS LATTICE OF THE UNIT CELL IN THE CRYSTAL

The location of each molecule in the lattice of a crystal is a *lattice point*. Although a lattice is a very orderly array, in some crystals the molecules are not all in the same orientation. Another issue is that the positions of the molecules in the crystal relative to one another vary in different types of crystals. To solve a molecular structure, the characteristics of the *unit cell* within the lattice must be specified (Glazer, 2016; Julian, 2015; Laudise, 1970; Massa, 2004; Stout & Jensen, 1989). A unit cell is a volume that has 6 or 8 sides, and 8 or 12 corners. There is a lattice point at each corner. The unit cell is the smallest volume that when repeated over and over in three dimensions can create the entire lattice of the crystal. For crystals of organic metabolites, the edges of a unit cell are typically between 3.0 and 40.0 Å long. The number of atoms per cell depends on the structure of the molecule in the crystal; however, it is usually not more than 1000.

If only the shapes of the unit cells are considered, the various types may be divided into seven *lattice systems*. The seven systems are: triclinic, monoclinic, orthorhombic, tetragonal, trigonal, hexagonal, and cubic. If the positions of the lattice points in the cells are considered one finds that a cell may have points only at the corners. This is a *primitive* unit cell. If lattice points are present on one or more surfaces of the cell and/or inside the cell, it is a *nonprimitive* unit cell.

A scientist named Bravais discovered that there are six types of primitive and eight types of nonprimitive unit cells in crystals. These arrangements are known as the 14 *Bravais lattices*.

8.6 IDENTIFYING THE SPACE GROUP OF THE UNIT CELL IN THE CRYSTAL

After the unit cell of a crystal is placed in a lattice system and a Bravais lattice, the *space group* must be identified. In addition to the shape of the unit cell and the positions of the molecules in each cell, the space group specifies the relative orientation of the molecules. There are 230 possible space groups. It may appear that assigning the space group is extremely challenging; however, a survey of 29,059 structures of organic molecules that were solved by crystallography revealed that 74.6% were classified in one of five of these space groups (Mighell, Himes, & Rodgers, 1983; Stout & Jensen, 1989).

8.7 RECORDING THE DIFFRACTION DATA AS A RECIPROCAL LATTICE OF THE CRYSTAL

The diffraction data are stored by computational software as a *reciprocal lattice* that is derived from the *direct lattice* of the crystal as follows. One corner of the unit cell is chosen as the origin and the three edges that extend from this corner are defined as the *basis vectors* or *crystallographic axes*: *a*, *b*, and *c*. The angles between these axes are symbolized as α, β, and γ. A *direct lattice* is specified by mathematical equations that give the positions of lattice points in terms of these six variables. As mentioned earlier in this chapter, in the type of crystals discussed here, each lattice point is one organic molecule.

The reflection planes, that are parallel to one another and that include all of the atoms in all of the organic molecules in the crystal, will intersect the crystallographic axes of the unit cells. Because the orientation of these planes relative to the unit cells changes as the crystal is rotated, the sites of intercept of the planes on the axes will also change. Each site of intercept is recorded as a fraction of the length of one of the three axes. Reciprocals of these fractions are the *Miller indices* that are symbolized as *h*, *k*, and *l*. The symbol *h* refers to the intercept on the *a* axis, *k* to the intercept on the *b* axis, and *l* to the intercept on the *c* axis. This specification of the orientation of the reflection planes creates the reciprocal lattice.

For example, at a particular moment during the collection of diffraction data, a reflection pane might intersect at 1/2 on the *a* axis, 1/3 on the *b* axis, and at 1/4 on the *c* axis. In this case the Miller index *h* is 2, *k* is 3, and *l* is 4. Because the sites of intercept change as the crystal is rotated, values of the Miller indices vary with time. The raw diffraction data that describe the reciprocal lattice are recorded by computational software as an *hkl file*.

8.8 SOLVING THE STRUCTURE OF AN ORGANIC METABOLITE

A first approximation to the molecular structure is done by interpretation of the *hkl* file with computational software in the WinGX package. WinGX may be downloaded at no cost from a website maintained by Louis Farrugia in the School of Chemistry, University of Glasgow, Scotland (Farrugia, 2012). Further refinement is performed with the SHELX-2017 computational software that may be downloaded at no cost from a website maintained by George Sheldrick in the Institute of Inorganic Chemistry, University of Göttingen, Germany (Sheldrick, 2008, 2010, 2015a, 2015b). Spatial variation in atomic vibrations is determined with the *Oak Ridge Thermal Ellipsoid Prediction* software that is part of the WinGX package. Imaging of the arrangement of molecules in the unit cell is done with the *Mercury* software that may be downloaded from the website of the CCDC (Cambridge_Crystallographic_ Data_Centre, 2018).

BIBLIOGRAPHY

Cambridge_Crystallographic_Data_Centre. (2018). *Homepage of the Cambridge Structural Database.* Cambridge, England. https://www.ccdc. cam.ac.uk.

Farrugia, L. J. (2012). WinGX and ORTEP for windows: An update. *Journal of Applied Crystallography, 45,* 849−854. Available from https://onlinelibrary. wiley.com/doi/abs/10.1107/S0021889812029111.

Glazer, A. M. (2016). *Crystallography: A very short introduction* (1st ed.). New York, NY: Oxford University Press. Available from https://lccn.loc.gov/ 2015958968.

Julian, M. M. (2015). *Foundations of crystallography with computer applications* (2nd ed.). Boca Raton, FL: CRC Press, Taylor & Francis Group. Available from https://lccn.loc.gov/2014006992.

Laudise, R. A. (1970). *The growth of single crystals.* Englewood Cliffs, NJ: Prentice-Hall. Available from https://lccn.loc.gov/77104173.

Massa, W. (2004). *Crystal structure determination* (2nd completely updated ed.). Berlin; New York, NY: Springer. Available from https://lccn.loc.gov/ 2003069465.

Mighell, A. D., Himes, V. L., & Rodgers, J. R. (1983). *Space-group frequencies for organic compounds. Acta Crystallographica Section A* (39, pp. 737−740). Available from https://onlinelibrary.wiley.com/doi/abs/10.1107/S01087673 83001464.

Nikon_Instruments_Inc. (2017). *Polarized light microscopy.* Melville, NY. https://www.microscopyu.com/techniques/polarized-light.

Sheldrick, G. M. (2008). A short history of SHELX. *Acta Crystallographica Section A: Foundations and Advances, 64,* 112−122. Available from https:// doi.org/10.1107/S0108767307043930.

Sheldrick, G. M. (2010). Experimental phasing with SHELXC/D/E: Combining chain tracing with density modification. *Acta Crystallographica Section D: Biological Crystallography, 66,* 479−485. Available from http://www.ncbi. nlm.nih.gov/pmc/articles/PMC2852312/.

Sheldrick, G. M. (2015a). Crystal structure refinement with SHELXL. *Acta Crystallographica Section C: Structural Chemistry, 71,* 3−8. Available from http://www.ncbi.nlm.nih.gov/pmc/articles/PMC4294323/.

Sheldrick, G. M. (2015b). SHELXT − Integrated space-group and crystal- structure determination. *Acta Crystallographica Section A: Foundations and Advances, 71,* 3−8. Available from http://www.ncbi.nlm.nih.gov/pmc/arti- cles/PMC4283466/.

Stout, G. H., & Jensen, L. H. (1989). *X-ray structure determination: A practical guide* (2nd ed.). New York, NY: Wiley. Available from https://lccn.loc.gov/ 88027931.

Chapter 9

Exercises in purifying and characterizing a quorum-sensing signal

9.1 GROWTH OF A CULTURE OF THE LUMINESCENT MARINE BACTERIUM *VIBRIO FISCHERI*

Theoretical background

There is some variation in the scientific literature regarding the genus into which the bacterial species *fischeri* should be classified. In the first report of the purification of *Vibrio* autoinducer 1 (VAI-1), the bacterial source was described as *Photobacterium fischeri* (Eberhard et al., 1981).

Sixteen years later a laboratory manual that described recombinant DNA experiments with genes from *fischeri* stated "*Vibrio fischeri* (which has recently been reclassified as *Photobacterium fischeri*)" (Winfrey, Rott, & Wortman, 1997). I followed this trend and stated *Photobacterium* as the genus in some of my publications (Crowley, 2010, 2011). After these articles were published I noticed that most scientists were still placing *fischeri* in *Vibrio*, therefore I described this species as *V. fischeri* in my most recent publication (Crowley & Kyte, 2014).

In 2007 studies of the genes in *fischeri* that encode 16S ribosomal RNA were reported (Urbanczyk, Ast, Higgins, Carson, & Dunlap, 2007). These studies suggested that *fischeri* should not be in either *Vibrio* or *Photobacterium*. The authors proposed a new genus, *Aliivibrio*, that would include *fischeri* and some other species of bacteria. The species *fischeri* is currently placed in the genus *Aliivibrio* in at least one textbook (Tortora, Funke, & Case, 2016) and at least one database of taxonomy (NCBI_Taxonomy_Aliivibrio_fischeri, 2017).

Purification and Characterization of Secondary Metabolites.
DOI: https://doi.org/10.1016/B978-0-12-813942-4.00009-7
© 2020 Elsevier Inc. All rights reserved.

Not all of the scientists that study the *fischeri* bacterium have started to use *Aliivibrio* as the genus (Kimbrough & Stabb, 2015; Thompson et al., 2017). It is not yet clear if a consensus exists in the scientific community regarding *fischeri* being in the genus *Aliivibrio* and it will be impractical to change the name of the VAI-1 quorum signal. Because of these two issues, I will continue to state the name of the genus as *Vibrio* in this book.

Bioluminescence and quorum sensing have been studied in more than one strain of *V. fischeri*. Two of the best-characterized strains are symbiotic with the Japanese pinecone fish, *Monocentris japonicus*. They are known as MJ1 (Eberhard et al., 1981; Winfrey et al., 1997) and MJ11. The most convenient bacterium to use for the exercises in this chapter is one of these two strains.

Technical background

The natural environment of *V. fischeri* bacteria is the ocean. They should be incubated at the ambient temperature in a laboratory, approximately 22°C. They will *not* grow well at the higher temperature that is typically used for incubation of bacteria such as *Escherichia coli* whose natural environment is the body of a warm-blooded animal. The growth of *V. fischeri* is faster if the bacteria have access to molecular oxygen (O_2) from the atmosphere. The chemical reaction that generates luminescence in this bacterium also requires oxygen. The flask containing the culture should be rotated or agitated during the incubation to get oxygen into the medium.

Observation of the expected luminescence in colonies on solid medium or bacteria in liquid medium will verify that the proper species is being propagated. This may be observed with the unaided eye; however, there are two issues worth noting. The luminescence will only be apparent in a dark room. A room that lacks windows, leaks allowing incoming light or electrical equipment emitting light should be used. Once you are in the dark room with the culture, allow at least 5.0 min for your pupils to dilate (dark adaptation) to maximize your sensitivity to luminescence.

The cellular density of the culture may be monitored by spectrophotometry as explained in the earlier chapter that provides an overview of the methods for purification of metabolites. The transmittance of visible radiation with a wavelength of 600 nm, T_{600}, or the optical density at this wavelength, OD_{600}, may be used for this calculation.

Materials

Reagents

- Good *Vibrio* medium (GVM) for the culture of bacteria as described in Table 9.1 (Winfrey et al., 1997).
  The medium should be autoclaved to make it aseptic.
- Colonies of the bacterium *V. fischeri* MJ1 or MJ11 on a plate of solid GVM medium.

Supplies and equipment

- Inoculating loop.
- Bunsen burner.
- Cuvettes for assaying the turbidity of the culture using the spectrophotometer.
- Spectrophotometer.
- Rotating platform for agitating the flask containing the bacterial culture during the incubation.
- A room that may be made completely dark for viewing luminescence.

TABLE 9.1 Good *Vibrio* medium for growing a culture of *V. fischeri* MJ1 or MJ11.[a]

	Grams per liter	Millimoles per liter
Tryptone	10.0	Not applicable, complex mixture
Casamino acids[b]	5.0	36.6
NaCl	25.0	430.0
$MgCl_2$	4.0	42.0
KCl	1.0	13.0
(Agar)[c]	(15.0)	Not applicable, complex mixture

[a]*The pH should be adjusted to 7.4 with NaOH.*
[b]*Casamino acids are generated by acid hydrolysis of the protein casein, a component of bovine milk. The molarity of the amino acids is calculated with the average molar mass of 136.8 g mol^{-1} and the assumption that each of the twenty amino acids is in equal abundance.*
[c]*Agar is only needed to prepare solid medium.*

Procedure

Grow the small starter culture

- Verify that the colonies on the solid medium are *V. fischeri* by looking for luminescence. The instructions for this assay were given earlier in *Technical background*.
- Inoculate a starter culture of GVM, between 5.0 and 20.0 mL, with a colony of *V. fischeri* MJ1 or MJ11.
- Incubate at ambient temperature with rotation at 90.0 revolutions per minute (rpm).
- After approximately 20.0 h of incubation, the culture should have dense growth. Look for the luminescence to verify that the microbes are *V. fischeri* rather than a contaminant. The cellular density of the starter culture should be approximately 1.0×10^8 mL^{-1}. This may be verified by spectrophotometry as explained earlier in *Technical background*.

Grow the large subculture

- Inoculate a subculture of 500.0 mL of GVM with 1.0 mL of the starter culture and incubate as before.
- After approximately 20.0 h of incubation, the culture should have dense growth. Look for the luminescence again to verify that the proper bacteria are growing. The cellular density of the subculture should be approximately 1.0×10^8 mL^{-1}. A precise measurement of the cellular density should be made and the total number of bacteria calculated.

9.2 REMOVAL OF BACTERIA FROM THE CULTURE BY CENTRIFUGATION AND FILTRATION

Theoretical and technical background

For the centrifugation described here, a *swinging-bucket rotor* is used. An optimal value of the *relative centrifugal force* (RCF) that is required is specified in this procedure. The definitions of swinging-bucket rotor and RCF are provided in the earlier chapter that gives an overview of the methods for purification of metabolites.

Materials

Reagents

- The culture of *V. fischeri* MJ1 or MJ11 (500.0 mL) from the previous exercise.

Supplies and equipment

- Bottles for centrifugation of the supernate. Two identical bottles, each of which has a capacity of at least 250.0 mL.
- A centrifuge and rotor combination with the following characteristics:
  1. The rotor should have swinging buckets that will hold at least two of the bottles described above.
  2. The dimensions of the rotor and the allowed rpm should be such that an RCF of $\sim 4000.0 \times g$ at the bottom of each bucket may be generated.
- A beaker or flask with a volume of at least 500.0 mL to hold the supernate between the centrifugation and filtration steps.
- Circular filters, made of glass fiber, purchased from Fisher Scientific. These filters should be of the G4 grade, have a retention of 1.2 μm and a diameter of 5.5 cm. Similar filters from a different supplier may be used as an alternative.
- A Buchner funnel with an internal diameter of ~ 5.5 cm.
- A filtering flask with a side arm that is constructed of thicker glass than is found in a typical flask. This flask will be used to create a vacuum underneath the Buchner funnel. The thick glass prevents the flask from collapsing. The volume should be at least 1.0 L.
- Plastic tubing that will not collapse when a vacuum is applied and is of the appropriate dimensions to connect the side arm on the filtering flask to the pump.
- A vacuum pump.
- Bottles for storage of the clarified supernate until the next exercise is performed. They should be constructed of plastic and have screw caps. If the supernate will be stored in a freezer, the volume of the bottles, and the material from which they are constructed, should be chosen to prevent cracking of the plastic.

Procedure

- Divide the culture into two equal portions and pour each portion into one of the plastic bottles that are designed for centrifugation.
- Measure the mass of each bottle. If the two masses differ by more than 0.1 g, adjust the volume of supernate in each so that the masses are within this range.

- Perform centrifugation so that the RCF at the bottom of each bottle is $\sim 4000.0 \times g$ for 60.0 min at ambient temperature (22.0°C).
- If it appears that greater than 90.0% of the bacteria have been sedimented, combine the two supernates by decanting them into a clean container.
- Perform vacuum-driven filtration to remove any remaining bacteria from the supernate. Use the filter, Buchner funnel, filtering flask, tubing, and pump that were described earlier in *Materials*. If you notice deterioration of the filter during this process, replace it with a fresh filter.
- The filtrate should appear pale brown but lack turbidity.
- Pour the clarified supernate into the bottles that are designed for storage. If the next exercise will not be performed until a later date, the containers should be shielded from light to prevent photolysis of VAI-1. Your instructor will decide whether to store the containers at ambient temperature, in a refrigerator or in a freezer.

Evaluation of results and questions to consider

- Comment on the effectiveness of centrifugation for sedimentation of the bacteria.
- Comment on whether or not the filtration step was necessary and whether or not it decreased the turbidity of the supernate.

9.3 EXTRACTION OF *VIBRIO* AUTOINDUCER 1, AN *N*-ACYL HOMOSERINE LACTONE, FROM THE SUPERNATE OF THE CULTURE WITH AN ORGANIC SOLVENT

Theoretical and technical background

Fundamental aspects of *liquid–liquid extraction*, also known as *solvent extraction*, are discussed in the earlier chapter in this book that provides an overview of the methods for purification of metabolites. The organic solvent ethyl acetate is immiscible with water. Although VAI-1 is soluble in water, it is more soluble in ethyl acetate. This metabolite may be extracted from the aqueous supernate of a bacterial culture into ethyl acetate.

Materials

Reagents

- The clarified bacterial supernate from the previous exercise, a volume of ~ 500.0 mL.
- Ethyl acetate containing 0.1% acetic acid and having a pH of ~ 5.0. A volume sufficient for two extractions of the aqueous supernate is needed, that is, ~ 1.0 L.

Supplies and equipment

- A separatory funnel made of glass, or other material that is resistant to ethyl acetate, with a volume of 500.0 mL.
- A ring stand to hold the separatory funnel.
- A beaker or flask, made of glass or other material that is resistant to ethyl acetate, and of sufficient size to hold the combined volume of the organic extracts (~ 1.0 L).

Procedure

Note: The supernate from the bacterial culture is an aqueous solution that includes VAI-1. The density of this solution is expected to be very close to that of pure water. Because the density of ethyl acetate (0.90 g mL^{-1}) is less than that of water, this organic solvent will be the upper layer in the separation step of each extraction.

- If the clarified supernate was frozen at the end of the previous exercise, thaw it in a bath of water maintained at $37°$C.
- Divide the supernate into two portions of ~ 250.0 mL each.
- Pour one portion of the aqueous supernate into the separatory funnel and extract it with an equal volume of the mixture of ethyl acetate and acetic acid. Recover the organic phase, which should now include the VAI-1, in the beaker or flask.
- Repeat the extraction of the first portion of the supernate and combine this organic extract with the first extract.
- Discard the first portion of the aqueous supernate.
- Pour the second portion of the aqueous supernate into the separatory funnel and then extract it twice with the mixture of organic solvents as was done with the first portion. Collect the two organic extracts of the second portion of the supernate and combine them with the two organic extracts of the first portion of the supernate.

- Discard the second portion of the aqueous supernate.
- If the next exercise will not be performed immediately, cover the container that holds the combined organic extracts and arrange shielding to prevent exposure to light that might cause photolysis of VAI-1. Store the container in a fume hood at ambient temperature (22.0°C).

9.4   REMOVAL OF SOLVENT FROM THE FRACTION ENRICHED IN *VIBRIO* AUTOINDUCER 1 AND RESUSPENSION OF THE DRIED SUBSTANCE

Theoretical and technical background

There are two parts to the process of drying of the solution that contains VAI-1. In the first part, water is removed by adsorption. In the second part, the organic solvent is removed by evaporation. Fundamental aspects of these techniques are presented in the earlier chapter in this book that provides an overview of the methods for purification of metabolites.

Materials

Reagents

- The organic liquid extract, which contains VAI-1, from the previous exercise. A volume of ∼1.0 L.
- Anhydrous magnesium sulfate, 5.0 g.
- Ethyl acetate for resuspension of the dry VAI-1, a volume of 8.0 mL.

Supplies and equipment

- A beaker with dimensions that will allow for rapid evaporation of solvent. If a beaker with a volume of 4.0 L is available, this will be ideal. The diameter of the opening at the top of a beaker is the same as the diameter of the base. Do not use a container in which the opening at the top has a smaller diameter than the diameter of the base (e.g., an Erlenmeyer flask). Evaporation is very slow in this type of flask.
- Beakers with a smaller volume for the final stage of evaporation, for example, 500.0 and 50.0 mL.
- A fume hood.

Procedure

- If water is evident underneath the organic liquid, estimate the volume. Add 0.5 g of anhydrous magnesium sulfate for every 1.0 mL of water.
- After the water has been adsorbed, decant the organic liquid into the beaker that will be used for evaporation of solvent. Avoid transferring the hydrated magnesium sulfate.
- Place the beaker that contains the organic liquid extract in a fume hood and leave it uncovered to allow the solvent, ethyl acetate, to evaporate.
- When the volume of solvent has decreased to $\sim$200.0 mL, pour the liquid into a beaker that has a volume of 500.0 mL. Allow the evaporation to continue in this smaller, uncovered beaker.
- When the volume of solvent has decreased to $\sim$20.0 mL, pour the liquid into a beaker that has a volume of 50.0 mL. Allow the evaporation to continue in this even smaller, uncovered beaker.
- After the evaporation has proceeded for a total of $\sim$50.0 h, no solvent should remain. The residue in the bottom of the beaker should be pale yellow.
- Resuspend the dry residue, which should include VAI-1, in 8.0 mL of ethyl acetate.
- Divide the sample into several portions. The portions should be stored in tightly sealed test tubes, shielded from light that might cause photolysis of this metabolite. If a freezer is available, this is the ideal place to store these samples.

9.5   ASSAY OF THE LUMINESCENCE-INDUCING ACTIVITY OF *VIBRIO* AUTOINDUCER 1

Theoretical background

The *lux* operon of *V. fischeri* includes *luxI*, a gene involved in synthesis of VAI-1. This operon also includes the genes required for response to the VAI-1 signal and the chemical reaction that generates luminescence. A *plasmid* is a circular, double-stranded DNA that replicates autonomously in a bacterium. The techniques of recombinant DNA technology were used to insert a copy of the *lux* operon that has a mutation in *luxI* into a plasmid (Pearson et al., 1994). This plasmid, pHV200I$^-$, also includes a gene that makes the host bacterium resistant to the antimicrobial chemical ampicillin. The strain of

E. coli that harbors this plasmid is known as VJS533 (pHV200I$^-$). It is a *reporter strain* for the luminescence-inducing activity of VAI-1.

Addition of VAI-1 to a culture of the reporter strain causes activation of expression of the *lux* operon and the bacteria then emit luminescence. The mutation in *luxI* prevents the host bacterium from synthesizing VAI-1, thus activation does not occur before addition of VAI-1 to the culture.

Technical background

The method for assay of the biological activity of VAI-1 has been published (Pearson et al., 1994; Schaefer, Hanzelka, Parsek, & Greenberg, 2000). The reporter strain grows well at 30.0°C. The ambient temperature in a typical laboratory, 22.0°C, is closer to the natural environment of *V. fischeri* and is best for the assay of the activity of VAI-1.

The cellular density of the cultures may be monitored by spectrophotometry as explained in the chapter earlier in this book that provides an overview of the methods for purification of metabolites. The transmittance of visible radiation with a wavelength of 600 nm, T_{600}, or the optical density at this wavelength, OD_{600}, may be used for this calculation.

Materials

Reagents

- A sample of the substance that has been purified from the culture of *V. fischeri*. It should be dissolved in ethyl acetate. If the procedures described in the earlier sections of this chapter were performed properly, a volume of 0.02 mL will be required for each assay. This should contain the VAI-1 that was secreted by approximately 1.0×10^8 bacteria.
- Luria−Bertani (LB) broth with ampicillin for the starter bacterial culture.

    Each liter of LB broth should contain:

    10.0 g tryptone

    5.0 g yeast extract

    10.0 g NaCl

    The pH should be adjusted to 7.2 and the medium should be made aseptic.

    Aseptic ampicillin should be added to a concentration of $100.0\ \mu g\ mL^{-1}$ just prior to inoculation.

TABLE 9.2 Luminescence assay medium for detection of the activity of VAI-1.[a]

	Grams per liter	Millimoles per liter
Tryptone	0.50	Not applicable, complex mixture
Glycerol	0.30	3.3
NaCl	5.8	100.0
$MgSO_4$	6.0	50.0
KH_2PO_4	0.68	5.0
K_2HPO_4	0.87	5.0
Ampicillin[b]	0.10	0.30

[a]The pH should be adjusted to 7.0.
[b]Prepare an aseptic stock solution of ampicillin and then add an appropriate aliquot to the medium when it is at ambient temperature.

- Luminescence assay medium (LAM).
  The components are listed in Table 9.2. The medium should be made aseptic.
- Colonies of the reporter strain, *E. coli* VJS533 (pHV200I$^-$), on an LB/agar/ampicillin plate.

Supplies and equipment
- Inoculating loop.
- Bunsen burner.
- Cuvettes for assaying the turbidity of the culture using the spectrophotometer.
- Spectrophotometer.
- An incubator that is set at 30.0°C and contains a rotating platform for incubation of the starter culture.
- A rotating platform at ambient temperature, approximately 22.0°C, for incubation of the cultures in which luminescence will be assayed.
- A dark room for observing luminescence.

Procedure
Note: Use aseptic technique for all manipulations.

Starter culture of the reporter strain

- Inoculate 5.0 mL of LB broth that contains ampicillin at 100.0 µg mL^{-1} with a colony of VJS533 (pHV200I$^-$).
- Incubate the starter culture at 30.0°C with rotation at 60.0 rpm for 12.0 h. At the end of this incubation, the cellular density should be approximately 4.0×10^8 cells mL^{-1}.

Subcultures of the reporter strain for the assay of luminescence

- Dilute 0.27 mL of the starter culture into 4.0 mL of sterile LAM. This is a 1:15 dilution.
- Split the inoculated luminescence assay culture into two sterile test tubes, with 2.0 mL of culture in each. The cellular density should be approximately 2.5×10^7 mL^{-1}.
- To one of the assay cultures add 0.02 mL of the purified substance. This is the *experimental culture*. To the other add 0.02 mL of ethyl acetate. This is the *negative control culture*.
- Incubate the cultures at ambient temperature, rotating at 120.0 rpm for 3.0 h. The bacteria will remain healthy during this incubation. Because ambient temperature is well below the optimal temperature for the growth of *E. coli*, there should not be a significant increase in the cellular density of the cultures during this incubation.
- To look for luminescence in the cultures, follow the instructions given in *Technical background* in the exercise in which the culture of *V. fischeri* was grown earlier in this chapter.

Evaluation of results and questions to consider

- Does the result of this assay convince you that VAI-1 has been purified?
- If no luminescence was observed, what are some of the possible explanations?
- If you are allowed to repeat the assay, what will be done differently?

9.6 SPECTROPHOTOMETRIC QUANTIFICATION OF THE YIELD OF *VIBRIO* AUTOINDUCER 1 USING ULTRAVIOLET RADIATION

Theoretical background

As is the case for most organic molecules that are not associated with a metallic ion, VAI-1 absorbs radiation with a wavelength in the near UV region but not visible radiation. The wavelength of maximum absorption (λ_{max}) for VAI-1 is 260 nm. The molar extinction coefficient, ε_{260}, is 2.5 mM^{-1} cm^{-1}.

Technical background

In the procedures in exercises presented earlier in this chapter, VAI-1 was extracted from the aqueous supernate of the bacterial culture into the organic solvent ethyl acetate. Ethyl acetate is much less polar than water. Although VAI-1 is soluble in the aqueous medium, it is extracted because it is more soluble in ethyl acetate.

To minimize the possibility that absorption of UV radiation by the ethyl acetate in the analyte will interfere with this assay, deionized water should be used as the solvent for the dilution that is described below in *Materials* and *Procedure*. Water does not absorb near UV radiation.

Materials

Reagents

- An aliquot of the VAI-1 that was purified from the supernate of a culture of *V. fischeri* and resuspended in ethyl acetate in earlier exercises. If your technical skills in the various steps of the purification were good, a sample of between 15.0 and 30.0 μL will be enough to acquire acceptable data. A greater volume will be needed if the yield was low. This aliquot of VAI-1 will be diluted into deionized water to provide sufficient volume for spectrophotometry. The design of the cuvette that is used will have an effect on the volume of sample that is necessary. If the volume of solution needed to fill the cuvette to the proper level is 0.5 mL, then 15.0 μL of sample should be enough. If 1.0 mL is required to fill the cuvette, then 30.0 μL of sample will be needed.
- Pure ethyl acetate to prepare a negative control to set the baseline for the spectrophotometric assay. A volume equal to the volume of VAI-1 that is diluted into water will be needed.

Supplies and equipment

- One or two rectangular cuvettes made of quartz that are transparent to near UV radiation. The volume of solution necessary to fill the cuvette to a level above the point at which the radiation will pass through should be determined before preparing the samples for the assay. For most cuvettes, 0.5 or 1.0 mL is sufficient. The distance through the solution in the cuvette that the radiation will pass through, known as the *length of the path*, should be measured. For most cuvettes, this is 1.0 cm. If a dual-beam spectrophotometer will be used, two cuvettes will be needed.

- A spectrophotometer that is capable of quantifying absorbance of radiation with wavelengths from 190 to 400 nm, that is, the near UV. It is best if the device will scan this range and plot the spectrum by means of associated software; however, if the only device available is one that records data at a fixed wavelength, it will be sufficient.

Procedure

Set the baseline

- Prepare a cuvette with the negative control solution. An aliquot of ethyl acetate should be diluted into deionized water to give a concentration of ethyl acetate equal to that in the diluted sample of VAI-1 that is described below.
- Set the baseline. If the device is capable of scanning, the baseline should be set for the range 190−400 nm. If the device only does fixed wavelengths, set it at 260 nm.

Prepare the sample of Vibrio *autoinducer 1 and quantify the absorbance*

- Dilute an aliquot of the preparation of VAI-1 into a volume of deionized water that is sufficient for the cuvette to be used. Refer to the discussion in *Materials* above regarding the appropriate volume of VAI-1 to be diluted.
- Record a spectrum of the entire range of near UV or the A_{260} of the diluted VAI-1. If the absorbance appears to be too weak to be accurate, prepare a new dilution with a larger aliquot of VAI-1 and repeat the spectrophotometric assay.

Evaluation of results and questions to consider

- If a scan of wavelengths was performed, was a distinct peak at 260 nm evident?
- Calculate the molar concentration of VAI-1 in the preparation prior to dilution for spectrophotometry. If your technique was good, millimolar will probably be appropriate. The following should be used.

1. The A_{260} of the diluted VAI-1.
2. The ε_{260} for VAI-1 that was stated earlier in this exercise.
3. The length of the path in the cuvette.
4. The Beer–Lambert law as described in the earlier chapter that provided an introduction to the use of UV radiation for spectrophotometry.
5. The dilution ratio for the sample in the cuvette.

- Using the total volume of the preparation of VAI-1, and the concentration you have just calculated, calculate the yield of VAI-1 in units of moles. If your technique was good, micromoles will probably be the most appropriate.
- Use 213.1 g mol^{-1} as the molar mass and calculate the yield of VAI-1 in units of mass.

9.7   INFRARED SPECTROSCOPIC ELUCIDATION OF THE STRUCTURE OF *VIBRIO* AUTOINDUCER 1

Theoretical background

A chapter presented earlier in this book explains how quantification of the absorption of infrared radiation by organic molecules is used to reveal their structure. The quantity that is recorded is the absorption of energy due to the stretching or bending of chemical bonds.

Technical background

Because the solvent in which VAI-1 is dissolved (ethyl acetate) is an organic molecule, it is imperative to distinguish absorption of energy by the bonds in the molecules of analyte from the absorption by bonds in the molecules of solvent. Some IR spectrometers are designed in a way that a solution of analyte may be allowed to dry, eliminating the solvent, before the spectrum is captured.

Materials

Reagents

- An aliquot containing 20.0 nmol of VAI-1 (in ethyl acetate) will be needed for each assay with the IR spectrometer. This corresponds to 4.2 μg. If your technical skills were good during the purification of VAI-1, a volume of approximately 2.0 μL should contain this amount.

Supplies and equipment

- Micropipettors and pipette tips.
- An IR spectrometer that is designed to assay organic substances.

Procedure

- If the device being used is designed to assay dry samples, place a drop containing 20.0 nmol of VAI-1 onto the sample-loading spot. Allow the solvent to evaporate.
- Acquire the spectrum of absorption of IR radiation.

Evaluation of results and questions to consider

- Compare the IR spectrum that you have recorded to the published spectrum for VAI-1 (Eberhard & Schineller, 2000; Eberhard et al., 1981). Is it possible that VAI-1 has been purified?

Note: In the article published in 1981, *Photobacterium* (rather than *Vibrio*) is stated as the genus for the bacterial species *fischeri*. This discrepancy is explained in an earlier section in this chapter.

9.8 ASSAY, BY MASS SPECTROMETRY, OF THE STRUCTURE OF *VIBRIO* AUTOINDUCER 1

Theoretical background

A chapter presented earlier in this book provided an introduction to mass spectrometry (MS). As explained in that chapter, a molecule must be in an ionic form to be projected though the electric field toward the detector. The uppercase letter "M" is used to represent a molecule of neutral analyte (Niessen, 2006).

For MS, there are several ways in which a neutral organic molecule may be converted to an ion. Two of these ways are the addition to, or loss of a proton from, this molecule. This generates a *protonated molecule* or a *deprotonated molecule* and this is specified with the notation +H or −H. The net charge on the ionic form is denoted by a plus or minus sign in superscript. The protonated form is thus denoted $[M + H]^+$, whereas $[M−H]^-$ represents the deprotonated form. To detect $[M + H]^+$ the mass spectrum is recorded in *positive-ion mode*, whereas *negative-ion mode* is used to detect $[M−H]^-$. VAI-1 may be detected as either an $[M + H]^+$ or an $[M−H]^-$ ion.

Technical background

For a pure substance, good data may be obtained by injecting it directly into an MS device. If the analyte is expected to be a mixture, or if it is possible that the purification method did not remove all of the undesired substances, the hybrid technique of liquid chromatography—mass spectrometry (LC-MS) should be used. If the chromatography parameters are set properly, LC will separate the substance of interest from other substances. In this way a mass spectrum of the pure substance of interest may be obtained.

As explained in the earlier chapter that was mentioned above, the data produced by MS is a *mass spectrum*. The symbol m represents the molar mass of an analyte in grams per mole and the symbol z represents the number of charges on each molecule of this analyte. A mass spectrum is a histogram of the relative amount of analyte molecules that have a particular m/z ratio.

LC—MS produces two types of data. For LC, a *chromatogram* in which the amount of analyte eluting from the column is plotted as a function of time is generated. For VAI-1, a mass detector will be used to generate the chromatogram. For analytes that efficiently absorb UV or visible radiation, spectrophotometry generates the chromatogram. If there is single peak on the chromatogram, a mass spectrum of the substance in this peak is examined. If there are multiple peaks, the mass spectrum of the peak that is believed to be the substance of interest is examined.

In the procedures in exercises presented earlier in this chapter, VAI-1 was extracted from the aqueous supernate of the bacterial culture into the organic solvent ethyl acetate. Ethyl acetate is much less polar than water. Although VAI-1 is soluble in the aqueous medium, it is extracted because it is more soluble in ethyl acetate.

The VAI-1 that you have purified is dissolved in ethyl acetate. You will probably need to dilute an aliquot of this preparation to decrease the concentration so that an appropriate amount will be injected into the device. Deionized water should be used as the solvent for the dilution. As explained below in *Materials*, detection of organic metabolites with MS is very sensitive. The dilution will probably be at least 100-fold and thus the concentration of ethyl acetate in the diluted solution will probably be less than 1.0%. It is unlikely that this amount of the organic solvent will interfere with either the chromatography or the spectrometry.

Materials

Reagents

- An aliquot of the VAI-1 that was purified, dried and then resuspended in ethyl acetate in earlier exercises. Because the amount of analyte necessary to obtain good data varies depending on the type of device employed and whether the MS or LC–MS technique is performed, your instructor and the laboratory staff will have to provide advice. A typical LC–MS assay of a bacterial secondary metabolite requires only 150.0 to 300.0 pmol of analyte. This corresponds to between 32.0 and 64.0 ng of VAI-1. Such an amount should be only a very small percentage of your preparation.

Supplies and equipment

- Micropipettors and pipette tips.
- Test tubes to prepare a dilution of the analyte.
- A device designed to assay the mass spectrum of an organic substance that is injected as a solution, or a hybrid device that performs liquid chromatographic separation of organic substances in solution and then injects the pure substances into a mass spectrometer.

Procedure

- A typical volume of analyte that is injected into an MS or LC–MS device is 5.0 µL. To prepare a sample with between 150.0 and 300.0 pmol of VAI-1 in this volume, it will probably be necessary to dilute an aliquot of the preparation into deionized water. Prepare the appropriate dilution.
- Inject the appropriate volume of diluted VAI-1 into the device and acquire the data.

Evaluation of results and questions to consider

- Compare the data that have been obtained to the published mass spectral data for VAI-1 (Eberhard & Schineller, 2000; Eberhard et al., 1981). Does your data suggest that VAI-1 has been purified?

Note: In the article published in 1981, *Photobacterium* (rather than *Vibrio*) is stated as the genus for the bacterial species *fischeri*. This discrepancy is explained in an earlier section in this chapter.

- What is the molecular formula of VAI-1? Is your data consistent with this formula?
- If LC–MS was performed and multiple peaks were observed on the chromatogram, are you convinced that the mass spectrum you examined was from the peak containing VAI-1? Explain your reasoning.

9.9 NUCLEAR MAGNETIC RESONANCE SPECTROSCOPIC ELUCIDATION OF THE STRUCTURE OF *VIBRIO* AUTOINDUCER 1

Theoretical background

As explained in the chapter presented earlier in this book in which this technique was introduced, there are several isotopes to which NMR spectroscopy may be applied. In organic substances the most abundant of these is the predominant isotope of hydrogen, ^1H. In this exercise, ^1H NMR spectroscopy of the analyte, VAI-1 that was purified in earlier exercises, is performed. This technique is also known as *proton NMR*.

Technical background

In the procedures in exercises presented earlier in this chapter, VAI-1 was extracted from the aqueous supernate of the bacterial culture into the organic solvent ethyl acetate. Ethyl acetate is much less polar than water. Although VAI-1 is soluble in the aqueous medium, it is extracted because it is more soluble in ethyl acetate.

Refer to the chapter presented earlier in this book in which NMR spectroscopy was introduced and to some of the cited references (Richards & Hollerton, 2011) for a discussion of appropriate solvents. One of the goals in choosing a solvent is minimization of resonance peaks in the spectrum due to the molecules of solvent. For ^1H NMR spectroscopy, the analyte, VAI-1 in this case, should be dissolved in a solvent in which the atoms of ^1H have been replaced with ^2H (deuterium). This is known as a *deuterated solvent* and it is used because ^2H does not generate a peak of resonance on the spectrum in this type of experiment. Deuterated versions of water or methanol are often used as solvents for organic metabolites being examined with NMR spectroscopy. The disadvantage of these two solvents is that exchange between the ^1H in acidic groups in the analyte and the ^2H in the solvent results in loss of the expected peaks of resonance for these protons.

The solvents most commonly used for NMR spectroscopy of VAI-1 are deutero-water (D_2O) and deutero-dimethyl sulfoxide (D_6-DMSO) (Eberhard et al., 1981). Because of the exchange of ^1H from acidic groups in molecules of analyte with ^2H in D_2O that was mentioned earlier, D_6-DMSO is the preferred solvent for this experiment.

Peaks of resonance due to the solvent are not completely eliminated by the use of ^2H. As explained in the earlier chapter mentioned above, some exchange of ^1H in the analyte with ^2H in the molecules of solvent will result in a solvent-specific peak or cluster of peaks in the NMR spectrum and this may complicate interpretation of the data. In an experiment in which an analyte is dissolved in D_6-DMSO be aware that under atmospheric pressure, the boiling point of this liquid is 189.0°C. After a solution in D_6-DMSO is prepared, it is difficult to increase the concentration of the solute because this solvent does not readily evaporate.

Materials

Reagents

- D_6-DMSO to dissolve the analyte, VAI-1, as explained in _Technical background_ earlier in this exercise. As discussed below, each assay usually requires less than 1.0 mL of solvent.
- An aliquot of the VAI-1 that was purified, dried and then resuspended in ethyl acetate in earlier exercises. If the frequency provided by the NMR spectrometer is 500.0 MHz or greater, 6.0 μmol (1.28 mg) of VAI-1 should be sufficient to obtain good data. If the VAI-1 is resuspended in D_6-DMSO at a concentration of 10.0 mM (2.13 mg mL^{-1}), then 0.6 mL will contain this amount of VAI-1. This is a typical volume for an NMR spectroscopic experiment.

Equipment and supplies

- Micropipettors and pipette tips.
- A test tube for evaporation of the ethyl acetate solvent from the sample of analyte and resuspension in D_6-DMSO.
- A fume hood or a centrifuge in a low-pressure chamber for evaporation of ethyl acetate.
- An _NMR sample tube_, that is, cylindrical glass tube designed to hold a solution of analyte that will be assayed with NMR spectroscopy. A tube with an outside diameter of 5.0 mm and a length of 18.0 cm is appropriate for a sample of 0.6 mL.
- An NMR spectrometer.

Procedure

- Consult with the instructor and/or the member of the laboratory staff that maintains the NMR spectrometer regarding the amount of analyte that will be necessary to obtain a good spectrum. Determine the number of moles necessary and the appropriate volume for the sample.
- Measure the appropriate volume of the VAI-1 solution (in ethyl acetate) into a tube that will allow for efficient evaporation of solvent.
- Leave the tube uncovered in a fume hood or use a centrifuge in an low-pressure chamber to remove the solvent by means of evaporation.
- Resuspend the analyte in D_6-DMSO and then use a micropipettor to transfer the sample into the NMR tube.
- Load the NMR tube that contains the analyte into the spectrometer and collect the data.

Evaluation of results and questions to consider

- Compare your results to the published NMR spectrum of VAI-1 (Eberhard & Schineller, 2000; Eberhard et al., 1981). Are the data from your experiment consistent with the conclusion that VAI-1 has been purified? Explain your reasoning.

Note: In the article published in 1981, *Photobacterium* (rather than *Vibrio*) is stated as the genus for the bacterial species *fischeri*. This discrepancy is explained in an earlier section in this chapter.

BIBLIOGRAPHY

Crowley, T. E. (2010). Expression, purification, and characterization of a recombinant flavin reductase from the luminescent marine bacterium Photobacterium leiognathi: A set of exercises for students. *Biochemistry and Molecular Biology Education*, *38*, 151–160. Available from https://www.ncbi.nlm.nih.gov/pubmed/21567817.

Crowley, T. E. (2011). Fluorescence-PCR assays and isolation of luminescent bacterial clones using an automated plate reader. *Biochemistry and Molecular Biology Education*, *39*, 126–132. Available from https://www.ncbi.nlm.nih.gov/pubmed/21445904.

Crowley, T. E., & Kyte, J. (2014). *Experiments in the purification and characterization of enzymes: A laboratory manual*. Amsterdam; Boston, MA: Elsevier/AP, Academic Press is an imprint of Elsevier. Available from https://lccn.loc.gov/2014395777.

Eberhard, A., Burlingame, A. L., Eberhard, C., Kenyon, G. L., Nealson, K. H., & Oppenheimer, N. J. (1981). Structural identification of autoinducer of *Photobacterium fischeri* luciferase. *Biochemistry, 20*, 2444–2449. Available from https://www.ncbi.nlm.nih.gov/pubmed/7236614.

Eberhard, A., & Schineller, J. (2000). Chemical synthesis of bacterial autoinducers and analogs. *Methods in Enzymology, 305*, 301–315. Available from http://www.ncbi.nlm.nih.gov/entrez/query.fcgi? cmd = Retrieve&db = PubMed&dopt = Citation&list_uids = 10812609.

Kimbrough, J. H., & Stabb, E. V. (2015). Antisocial luxO mutants provide a stationary-phase survival advantage in *Vibrio fischeri* ES114. *Journal of Bacteriology, 198*, 673–687. Available from https://www.ncbi.nlm.nih.gov/ pubmed/26644435.

NCBI_Taxonomy_Aliivibrio_fischeri. (2017). *Aliivibrio fischeri. Taxonomy ID 668, the bacterial species fischeri*. NCBI_Taxonomy_Database. https:// www.ncbi.nlm.nih.gov/Taxonomy/Browser/wwwtax.cgi?mode = Info&id = 668&lvl = 3&lin = f&keep = 1&srchmode = 1&unlock.

Niessen, W. M. A. (2006). *Liquid chromatography–mass spectrometry* (3rd ed.). Boca Raton, FL: CRC/Taylor & Francis. Available from https://lccn.loc.gov/ 2006013709.

Pearson, J. P., Gray, K. M., Passador, L., Tucker, K. D., Eberhard, A., Iglewski, B. H., & Greenberg, E. P. (1994). Structure of the autoinducer required for expression of Pseudomonas aeruginosa virulence genes. *Proceedings of the National Academy of Sciences of the United States of America, 91*, 197–201. Available from https://www.ncbi.nlm.nih.gov/pubmed/8278364.

Richards, S. A., & Hollerton, J. C. (2011). *Essential practical NMR for organic chemistry*. Chichester, West Sussex, UK: John Wiley. Available from https:// lccn.loc.gov/2010033319.

Schaefer, A., Hanzelka, B., Parsek, M., & Greenberg, E. (2000). Detection, purification, and structural elucidation of the acylhomoserine lactone inducer of *Vibrio fischeri* luminescence and other related molecules. *Methods in Enzymology, 305*, 288–301. Available from http://www.ncbi.nlm.nih.gov/entrez/query. fcgi?cmd = Retrieve&db = PubMed&dopt = Citation&list_uids = 10812608.

Thompson, L. R., Nikolakakis, K., Pan, S., Reed, J., Knight, R., & Ruby, E. G. (2017). Transcriptional characterization of *Vibrio fischeri* during colonization of juvenile Euprymna scolopes. *Environmental Microbiology, 19*, 1845–1856. Available from https://www.ncbi.nlm.nih.gov/pubmed/28152560.

Tortora, G. J., Funke, B. R., & Case, C. L. (2016). *Microbiology: An introduction* (12th ed.). Boston, MA: Pearson. Available from https://lccn.loc.gov/ 2014038680.

Urbanczyk, H., Ast, J. C., Higgins, M. J., Carson, J., & Dunlap, P. V. (2007). Reclassification of *Vibrio fischeri, Vibrio logei, Vibrio salmonicida* and *Vibrio wodanis* as *Aliivibrio fischeri* gen. nov., comb. nov., *Aliivibrio logei* comb. nov., *Aliivibrio salmonicida* comb. nov. and *Aliivibrio wodanis* comb. nov. *International Journal of Systematic and Evolutionary Microbiology, 57*, 2823–2829. Available from https://www.ncbi.nlm.nih.gov/pubmed/18048732.

Winfrey, M. R., Rott, M. A., & Wortman, A. T. (1997). *Unraveling DNA: Molecular biology for the laboratory*. Upper Saddle River, NJ: Prentice-Hall. Available from https://lccn.loc.gov/96036610.

Chapter 10

Exercises in purifying and characterizing iron-chelating molecules

10.1 GROWTH OF A CULTURE OF THE ENTERIC BACTERIUM *ENTEROBACTER AEROGENES*

Theoretical background

Aerobactin is a linear siderophore with three carboxylic acid groups and two hydroxamic acid groups. The molecular formula of the desferric form, that is, lacking the ion of iron(III), has been reported as $C_{22}H_{36}N_4O_{13}$ (Gibson & Magrath, 1969; Haygood, Holt, & Butler, 1993) and the molar mass has been reported as $564.24 \, g \, mol^{-1}$ (Haygood et al., 1993). Assuming that the average atomic mass of an ion of iron(III) is $55.84 \, g \, mol^{-1}$, the molar mass of ferric aerobactin that includes this metal ion is predicted to be $620.08 \, g \, mol^{-1}$.

Aerobactin was first purified from a bacterium classified as *Aerobacter aerogenes* (Gibson & Magrath, 1969). This species was later reclassified as *Klebsiella pneumoniae* (Haygood et al., 1993; Küpper, Carrano, Kuhn, & Butler, 2006; Pecoraro, Wong, Kent, & Raymond, 1983). Although the name of this sidero-phore appears to be derived from the name of the bacterial genus before the reclassification, that is, *Aerobacter*, the name aerobactin is still used.

Subsequent studies revealed that aerobactin is expressed by several other species of bacteria that are not in the genus *Klebsiella*. For example, secretion of this siderophore by *Enterobacter cloacae* has been detected (Keller, Pedroso, Ritchmann, & Silva, 1998; Van Tiel-Menkveld, Mentjox-Vervuurt, Oudega, & de Graaf, 1982). This observation suggests that aerobactin may be secreted by other species in the genus *Enterobacter*, such as *Enterobacter aerogenes*.

Purification and Characterization of Secondary Metabolites.
DOI: https://doi.org/10.1016/B978-0-12-813942-4.00010-3

The protocol described in this chapter for purification of a siderophore that has been secreted into the culture medium by *E. aerogenes* is similar to that used to purify aerobactin that was secreted from a bacterial species in the genus *Vibrio* (Haygood et al., 1993). If *E. aerogenes* expresses aerobactin, it is likely that this will be the siderophore that is purified. Even if aerobactin is not purified, this method will probably yield a siderophore with a similar structure.

Technical background

In the chapter that provides an overview of the methods for purification of metabolites that was presented earlier in this book, the difference between *complex media* and *chemically defined media* for the growth of bacterial cultures is discussed. Complex media contain many organic substances that may copurify with the secreted metabolite. The identity and abundance of the some of the organic substances in complex media is not known. Chemically defined media contain only one, or a small number, of organic substances and the concentration of each organic molecule is known.

E. aerogenes and *Escherichia coli* are both in the family Enterobacteriaceae (Johnson & Case, 2018; Tortora, Funke, & Case, 2019). Bacterial species in this family are commonly known as enterics because they inhabit the intestines of humans and other animals. *E. aerogenes* and *E. coli* are Gram-negative, rod-shaped, facultative anaerobes. In the procedure provided in this exercise, the culture medium, *glucose minimal-salts and amino acids broth* (GSAA broth), contains glucose and all 20 amino acids that are typically found in polypeptides. It does not contain iron. The recipe for this medium is derived from a medium known as *glucose minimal-salts broth* that is known to support the growth of *E. coli* (Johnson & Case, 2018; Tortora et al., 2019). Facultative anaerobes can proliferate in the absence of molecular oxygen (O_2); however, they grow faster in the presence of O_2.

E. aerogenes grows well in GSAA broth that is aerated by agitation and maintained at 37°C. In such a culture, these bacteria secrete a siderophore. The purity of the substance that is obtained from the procedures in the following exercises is sufficient for assays of chelation to prove that the substance is in fact a siderophore; however, refer to the note below regarding a culture medium that may provide a higher degree of purity.

The growth of the culture may be monitored by spectrophotometry as explained in the chapter that provides an overview of the methods for purification of metabolites earlier in this book. The transmittance of visible radiation with a wavelength of 600 nm, T_{600}, or the optical density at this wavelength, OD_{600}, may be used to quantify the turbidity. This quantity may then be used to calculate the cellular density of the culture.

Note: In the presence of an inorganic source of nitrogen, many species of bacteria can synthesize all 20 of the amino acids that are needed to make polypeptides. An inorganic salt is the only nitrogen-containing substance in the glucose minimal-salts broth mentioned previously for the growth of *E. coli*. Because *E. aerogenes* and *E. coli* are in the same family, it is likely that *E. aerogenes* will also grow in medium with inorganic nitrogen rather than a mixture of amino acids. An example of this is the Hoitink-Sinden optimized for production of coronatine (HSC) medium that is described in the chapter in this book in which the phytotoxin coronatine is purified. HSC medium includes 20.0 mM ammonium chloride and 3.0 mM potassium nitrate, but no organic substances that contain nitrogen. Your instructor may suggest that a medium similar to HSC be used to grow *E. aerogenes*. The lack of the amino acids in the supernate of the bacterial culture may make the purification of the siderophore more reliable. The 20.0 μM $FeCl_3$ that is a component of the HSC medium should not be included because the presence of iron(III) at this concentration will repress secretion of the siderophore from *E. aerogenes*.

Materials

Reagents
- Sterile medium for the bacterial cultures (Table 10.1). The volume of the starter culture will be 25.0 mL. The combined volume of the batches of subculture will be 1.0 L. In all cases use a flask that is large enough to allow the broth to slosh around and provide aeration when agitated.
- Colonies of *E. aerogenes* strain CECT 684 on a plate of nutrient agar (CECT is a registered trademark for the Spanish Type Culture Collection in Valencia, Spain).

Supplies and equipment
- Inoculating loop.
- Bunsen burner.
- Cuvettes for assaying the turbidity of the culture using the spectrophotometer.
- Spectrophotometer.

TABLE 10.1 Glucose, minimal-salts and amino acids broth for the culture of *Enterobacter aerogenes*.[a]

	Grams per liter	Millimoles per liter
Glucose	5.0	28.0
Casamino acids[b]	5.0	36.6
NaCl	5.0	86.0
$MgSO_4$	0.24	2.0
KH_2PO_4	1.0	7.5
K_2HPO_4	1.3	7.5

[a]*The pH of the broth should be adjusted to 6.6.*
[b]*Casamino acids are generated by acid hydrolysis of the protein casein, a component of bovine milk. The molarity of the amino acids is calculated with the average molar mass of 136.8 g mol^{-1} and the assumption that each of the 20 amino acids is in equal abundance.*

- Test tubes and flasks of the appropriate dimensions and shape for the growth of microbial cultures with aeration.
- A rotating platform or similar device at 37°C for agitation of the broth cultures to provide aeration.

Procedure

Grow the small starter culture

- Inoculate 25.0 mL of sterile GSAA broth with a colony of *E. aerogenes* CECT 684 bacteria.
- Incubate the bacterial culture at 37°C, rotating at 60 rpm for approximately 12.0 h. At the end of this period the cellular density of the culture should be between 1.0×10^8 and 1.0×10^9 mL^{-1}. Such a density means that the culture is in exponential phase. The density may be quantified as discussed earlier in this exercise in *Technical background*.

Grow the large subculture

- Obtain batches of sterile GSAA broth that have a combined volume of 1.0 L.
- Using aseptic technique, inoculate these batches of medium with a combined volume of 0.4 mL of the starter culture. This is a dilution ratio of 1:2500.

- Incubate the flasks of bacterial culture at 37°C, rotating at 60 rpm, for between 16.0 and 23.0 h. At the end of this period, the cellular density of the cultures should be approximately $2.0 \times 10^8 \, mL^{-1}$. Such a density means that the culture is in exponential phase. The density may be quantified as discussed earlier in this exercise in *Technical background*.

10.2 REMOVAL OF BACTERIA FROM THE CULTURE BY CENTRIFUGATION AND FILTRATION

Theoretical and technical background

For the centrifugation described here a *swinging-bucket rotor* is used. An optimal value of the *relative centrifugal force* (RCF) that is required is specified in this procedure. The definitions of swinging-bucket rotor and RCF are provided in the earlier chapter that gives an overview of the methods for purification of metabolites.

Although none of the components of the chemically defined medium in the culture are supposed to contain iron, there may be a low level of contamination in one or more of them. If some iron(III) is present, it is likely to be chelated by siderophores secreted by *E. aerogenes.* It is best to purify the desferric form of the siderophore, so that assays for chelation of iron(III) may be performed.

Chelation of iron(III) by siderophores has been observed to be more efficient in a solution with a pH near neutral rather than one with a low pH (Pecoraro, Harris, Wong, Carrano, & Raymond, 1983; Young & Gibson, 1979). For a siderophore containing hydroxamic acid groups, this may be explained by protonation at low pH that prevents the formation of ionic bonds with iron(III). The procedure below states that an aliquot of concentrated acid should be added to the clarified supernate to decrease the pH to ~ 2.3.

Materials

Reagents

- The cultures of *E. aerogenes*, a combined volume of 1.0 L, grown in the previous exercise.
- Aqueous hydrochloric acid at a concentration of 12.0 M.

Supplies and equipment

- Bottles for centrifugation of the supernate. Two or four identical bottles, with a combined volume of at least 1.0 L.
- A centrifuge and rotor combination with the following characteristics:
  1. The rotor should have swinging buckets that will hold the bottles described previously.
  2. The dimensions of the rotor and the allowed revolutions per minute should be such that a RCF of $\sim 4000.0 \times g$ at the bottom of each bucket may be generated.
- A beaker or flask with a volume of at least 1.0 L to hold the supernate between the centrifugation and filtration steps.
- Circular filters, made of glass fiber, purchased from Fisher Scientific. These filters should be of the G4 grade, have a retention of 1.2 μm and a diameter of 5.5 cm. Similar filters from a different supplier may be used as an alternative.
- A Buchner funnel with an internal diameter of ~ 5.5 cm.

A *filtering flask* with a side arm that is constructed of thicker glass than is found in a typical flask. This flask will be used to create a vacuum underneath the Buchner funnel. The thick glass prevents the flask from collapsing. The volume should be at least 1.0 L.

- Plastic tubing that will not collapse when a vacuum is applied and is of the appropriate dimensions to connect the side arm on the filtering flask to the pump.
- A vacuum pump.
- Bottles for storage of the clarified supernate until the next exercise is performed. They should be constructed of plastic and have screw caps. If the supernate will be stored in a freezer, the volume of the bottles, and the material from which they are constructed, should be chosen to prevent cracking of the plastic.

Procedure

Removal of the bacteria from the culture

- Divide the culture into two or four equal portions and pour each portion into one of the plastic bottles that are designed for centrifugation.
- Measure the mass of each bottle. If the masses differ by more than 0.1 g, adjust the volume of supernate in each so that the masses are within this range.

- Perform centrifugation so that the RCF at the bottom of each bottle is $\sim 4000.0 \times g$ for 60.0 min at ambient temperature (22.0°C).
- If it appears that greater than 90.0% of the bacteria have been sedimented, combine the supernates by decanting them into a clean container.
- Perform vacuum-driven filtration to remove any remaining bacteria from the supernate. Use the filter, Buchner funnel, filtering flask, tubing, and pump that were described earlier in *Materials*. If you notice deterioration of the filter during this process, replace it with a fresh filter.
- The filtrate should appear pale yellow but lack turbidity.

Acidification of the clarified supernate from the bacterial culture

- Assay the pH of the clarified supernate. It should be between 6.0 and 7.0.
- If the pH is in the expected range, the proper volume of 12.0 M HCl to add is 3.0 mL for the 1.0 L of supernate. After adding the acid, stir the supernate to equalize the pH throughout the solution.
- Assay the new pH. It should be approximately 2.3.
- Pour the acidified supernate into the bottles that are designed for storage. If the next exercise will not be performed until a later date, the containers should be shielded from light to prevent photolysis of the siderophore. Your instructor will decide whether to store the containers at ambient temperature, in a refrigerator or in a freezer.

Evaluation of results and questions to consider

- Comment on the effectiveness of centrifugation for sedimentation of the bacteria.
- Comment on whether or not the filtration step was necessary and whether or not it decreased the turbidity of the supernate.

10.3   EXTRACTION OF A SIDEROPHORE FROM THE SUPERNATE OF A CULTURE BY ADSORPTION ONTO A SOLID PHASE

Theoretical and technical background

The technique of *solid-phase extraction*, or SPE, is discussed in Section 3.5 in the earlier chapter that presents an overview of the methods for purification of metabolites. The solid phase that is the *adsorbent* used in this case is Amberlite XAD-2 from MilliporeSigma. This material consists of beads of a hydrophobic copolymer of styrene-divinylbenzene resin. A similar material from another supplier may be used as an alternative. In an aqueous solution such as a bacterial supernate, the siderophore from *E. aerogenes* will bind to this adsorbent. When the aqueous solution is replaced with methanol, which is less polar than water, the siderophore is eluted from the adsorbent.

Materials

Reagents

- The clarified and acidified supernate of the culture of *E. aerogenes* from the previous exercise. A volume of ~1.0 L.
- Beads of Amberlite XAD-2 from MilliporeSigma (dry), or a similar material from another supplier. A mass of 75.0 g is needed.
- Methanol. A volume that is sufficient to submerge the adsorbent is needed.

Supplies and equipment

- Two beakers or flasks. One for soaking of the adsorbent before it is mixed with the supernate. One for stirring the supernate into which the adsorbent has been dispersed.
- Magnetic stirring bar.
- A device for driving rotation of the stirring bar inside a beaker or flask.

Procedure

Preparation of the solid phase for the process of adsorption

- In a beaker or flask, soak the adsorbent in methanol for 15.0 min. The beads will increase in size in this solvent because it is less polar than water. This is known as *swelling* of the adsorbent.
- Decant and discard the excess methanol.
- Soak the adsorbent in deionized water for 5.0 min.
- Decant and discard the excess water.
- Repeat the wash with water.

Adsorption

- Decant the excess water from the adsorbent.
- In a beaker or flask, mix the adsorbent with the 1.0 L of clarified and acidified supernate.
- Stir the mixture for 3.0 h at ambient (22.0°C) temperature, to allow the adsorbent to capture the siderophore. Stirring will keep the beads of adsorbent suspended and dispersed, thus creating a *slurry*.
- If the next exercise will not be performed in the same session, cover the container that holds the adsorbent and arrange shielding from light to prevent photolysis of the siderophore.

10.4 ELUTION OF A SIDEROPHORE FROM A SOLID PHASE, REMOVAL OF SOLVENT AND RESUSPENSION OF DRIED SUBSTANCE

Theoretical and technical background

As is explained in the chapter in which an overview of the methods for purification of metabolites was presented, it is best to use a *chromatographic column* that contains the adsorbent and siderophore for the process of elution. This application is not chromatography; however, the use of such an apparatus produces an eluant in which the solvent is predominantly methanol. Minimization of the amount of water in the eluant is desired.

The pH of the solution in which the siderophore from *E. aerogenes* is dissolved has multiple effects on the properties of this molecule. In addition to the effect on chelation of iron(III) that was mentioned in an earlier exercise in this chapter, the spectrum of absorption of IR radiation is also affected by the pH. The latter effect is an issue in the exercise later in this chapter that involves IR radiation.

Addition of an aqueous solution of ammonium hydroxide, NH_4OH (aq), to the solution of substance that has been eluted from the solid phase in this procedure elicits a colorless precipitate. This precipitate does not contain the siderophore. If the precipitate is packed to the bottom of the test tube by centrifugation and the solution above it is transferred to a fresh tube, the transferred solution will be a more pure preparation of the siderophore. Ammonium hydroxide is a base. As discussed below, the pH of the clarified solution will be much higher than that of the solution prior to this step. The higher pH after the precipitation is desired for the assays in later exercises for chelation of iron(III).

Materials

Reagents

- The adsorbent from the previous exercise onto which the siderophore has been adsorbed.
- A volume of methanol ($\sim$100.0 mL) sufficient to elute the siderophore from the adsorbent.
- NH_4OH (aq) with a concentration of 9.0 M. A volume of 1.0 mL will probably be sufficient.

Supplies and equipment

- An apparatus for preparation of a chromatographic column that consists of the following. A cylindrical tube, made of glass or clear plastic that is resistant to methanol, mounted vertically on a ringstand. The tube should have a plug at the bottom in which a stopcock is inserted. Typical dimensions that are appropriate are a diameter of 5.0 cm and a height of 20.0 cm. Such a cylinder has a volume of 392.7 mL. The solvent to be used for elution, methanol, is volatile and hazardous. The apparatus should be assembled in a fume hood.
- Beakers of appropriate dimensions, as follows, for evaporation of the solvent from the solution eluted from the adsorbent. If the volume of this solution is $\sim$100.0 mL, the first beaker should have a volume of $\sim$1.0 L. As the volume of solvent decreases, beakers with volumes of $\sim$500.0 and $\sim$50.0 mL should be used. The diameter of the opening at the top of a beaker is the same as the diameter of the base. Do not use a container in which the opening at the top has a smaller diameter than the diameter of the base (e.g., an Erlenmeyer flask). Evaporation is very slow in this type of flask.
- A fume hood.
- A miniature centrifuge for centrifugation of test tubes that have a volume of between 1.5 and 7.0 mL.
- Indicator paper for assay of the pH of a solution.

Procedure

Removal of excess bacterial supernate from the adsorbent

- Halt the stirring of the mixture of bacterial supernate and adsorbent. Allow the beads to settle to the bottom of the container.
- Record the appearance of the solution above the beads. The appearance will be addressed in the evaluation section at the end of this exercise.

- Briefly stir the mixture of bacterial supernate and adsorbent to suspend the beads and create a slurry.
- Close the stopcock on the valve at the bottom of the cylindrical tube. Pour the slurry into the tube and allow the beads of adsorbent to settle to the bottom. This is now a chromatographic column.
- Open the stopcock and allow the aqueous liquid to drain out of the bottom of the column. When the meniscus reaches the top of the packed adsorbent, close the stopcock. The adsorbent should remain covered with the aqueous solution.

Elution of the siderophore from the adsorbent

Note: To release the siderophore, use 1.33 mL of methanol per gram of adsorbent. If the column holds 75.0 g of adsorbent, then 100.0 mL of methanol should be used.

- Pour the methanol into the column and then open the stopcock to begin the elution.
- Collect the eluant, ~100.0 mL, in a container made of glass or clear plastic that is resistant to methanol.
- When the meniscus of the methanol reaches the top of the packed adsorbent, close the stopcock to stop the elution.
- Record the appearance of the eluant. The appearance will be addressed in the evaluation section at the end of this exercise.

Removal of solvent by evaporation

Note: The solvent of the eluant should be at least 90.0% methanol. This organic liquid evaporates more rapidly than water. After all of the methanol has evaporated, some water (less than 10.0 mL) will remain in the beaker. This water will evaporate eventually but at a slower rate than observed for methanol.

- If the volume of the eluant is ~100.0 mL, pour it into a beaker with a volume of ~1.0 L.
- Leave the beaker uncovered in a fume hood to allow the solvent to evaporate.
- When the volume of solvent has decreased to ~50.0 mL, pour the liquid into a beaker that has a volume of ~500.0 mL. Allow the evaporation to continue in this smaller, uncovered beaker.
- When the volume of solvent has decreased to ~20.0 mL, pour the liquid into a beaker that has a volume of ~50.0 mL. Allow the evaporation to continue in this even smaller, uncovered beaker.
- After the total time of evaporation is ~48.0 h, all of the liquid should be gone. Record the appearance of the dry residue.

Resuspension of the dried substance in methanol

Note: The substance that has been purified may be aerobactin or a similar siderophore. The solubility of aerobactin in methanol is at least 5.7 mM, corresponding to 3.2 mg mL^{-1} (Harris, Carrano, & Raymond, 1979). If the volume of the culture of *E. aerogenes* from which the siderophore was extracted was 1.0 L, then 6.0 mL of methanol will be sufficient to dissolve the dry siderophore.

- Add 6.0 mL of methanol to the dry substance. Scrape and stir for a few minutes to dissolve the solid material. Some color will probably be observed in the liquid as the substance dissolves.
- If not all of the substance dissolves, transfer the solution to a test tube, leaving the undissolved material in the beaker. All of the siderophore should be dissolved in the methanol; therefore, the solid material in the beaker may be discarded.
- Record the appearance of the solution. The appearance will be addressed in the evaluation section at the end of this exercise.
- Apply a drop of the solution to indicator paper to determine the pH. It should be no higher than the pH recorded after acidification of the bacterial supernate in an earlier exercise.

Removal of undesired substances and increase of the pH by addition of ammonium hydroxide

- Add a volume of 9.0 M NH$_4$OH (aq) that is equal to one-ninth of the volume of the resuspended substance. Cap the test tube and invert to mix. The final concentration of NH$_4$OH should be 0.9 M and a colorless precipitate should form.
- Use centrifugation to pack the precipitate down to the bottom of the test tube. Transfer the clarified solution to a fresh test tube. Record the appearance of this solution. The appearance will be addressed in the evaluation section at the end of this exercise.
- Apply a drop of the clarified solution to indicator paper to determine the pH. It should be ∼8.0.
- Divide the solution of siderophore in methanol into several aliquots and tightly seal the test tubes. If a freezer is available, this is the ideal location for storage.

Evaluation of results and questions to consider

- Pure methanol has no color. Is a solution of a pure, desferric, hydroxamic acid-containing siderophore expected to have color? Explain your reasoning.
- Comment on the color of the preparation at the following steps in the purification. Discuss the implications of variation in color. Discuss how the change in color is correlated with an increase in the purity of the preparation.
  1. The bacterial supernate prior to SPE.
  2. The aqueous solution above the beads of adsorbent after 3.0 h of adsorption.
  3. The eluant from the adsorbent in which the solvent was predominantly methanol.
  4. The substance, which is expected to include the siderophore, that was resuspended in methanol after evaporation of solvent from the eluant.
  5. The solution after the precipitation that was caused by addition of NH_4OH (aq).

10.5 SPECTROPHOTOMETRIC QUANTIFICATION OF THE YIELD OF SIDEROPHORE USING ULTRAVIOLET RADIATION

Theoretical background

Because it contains no metallic ions, the desferric form of a siderophore is not expected to absorb visible radiation. Spectrophotometric quantification of the concentration of a desferric siderophore in a solution is performed with radiation in the near UV region.

The substance that has been purified may be aerobactin or a siderophore with a similar structure. A value for the ε_{500} of ferric aerobactin has been published (Gibson & Magrath, 1969). The A_{500} (visible radiation) of the ferric form of the siderophore from *E. aerogenes*, purified by the procedure described in this chapter and then mixed with ferric chloride, was quantified. The molar concentration of the siderophore in the original preparation, in the desferric form, was then calculated. The A_{280} of the desferric form was then quantified. The molar concentration, calculated from the A_{500} and ε_{500}, and the value of the A_{280} were then used to calculate that ε_{280} equals $18.8 \text{ mM}^{-1} \text{ cm}^{-1}$ for the desferric form of this siderophore.

Materials

Reagents

- An aliquot of the siderophore preparation that was eluted from the solid phase in methanol and then clarified by precipitation with ammonium hydroxide. If your technical skills in the various steps of the purification were good, a sample of 2.0 or 4.0 μL will be enough to acquire acceptable data. A greater volume will be needed if the yield was low. This aliquot of the siderophore sample will be diluted into deionized water to provide sufficient volume for spectrophotometry. The design of the cuvette that is used will have an effect on the volume of sample that is necessary. If the volume of solution needed to fill the cuvette to the proper level is 0.5 mL then 2.0 μL of sample should be enough. If 1.0 mL is required to fill the cuvette, then 4.0 μL of sample will be needed.

Supplies and equipment

- One or two rectangular cuvettes made of quartz that is transparent to near UV radiation. The volume of solution necessary to fill the cuvette to a level above the point at which the radiation will pass through should be determined before preparing the samples for the assay. For most cuvettes, 0.5 or 1.0 mL is sufficient. The distance through the solution in the cuvette that the radiation will pass through, known as the *path length*, should be measured. For most cuvettes this is 1.0 cm. If a dual-beam spectrophotometer will be used, two cuvettes will be needed.
- A spectrophotometer that is capable of quantifying absorbance of radiation with wavelengths from 190 to 400 nm, that is, the near UV. It is best if the device will scan this range and plot the spectrum by means of associated software; however, if the only device available is one that records data at a fixed wavelength, it will be sufficient.

Procedure

Set the baseline

- Prepare a cuvette with the appropriate blank solution. The siderophore is dissolved in methanol and will be diluted into deionized water; therefore, the blank should be prepared in the same manner, substituting pure methanol for the siderophore-in-methanol sample.

- Set the baseline. If the device is capable of scanning, the baseline should be set for the range 190−400 nm. If the device only does fixed wavelengths, set it at 280 nm.

Prepare the sample of siderophore and quantify the absorbance

- Dilute an aliquot of the siderophore sample into a volume of deionized water sufficient for the cuvette to be used. Refer to the discussion in *Materials* above regarding the appropriate volume of the siderophore to be diluted.
- Record a spectrum of the entire range of near UV or the A_{280} of the diluted siderophore. If the absorbance appears to be too weak to be accurate, repeat the assay with a larger aliquot of the siderophore sample. If a larger aliquot of the siderophore is used, then the setting of the baseline should be performed with the appropriate greater volume of methanol diluted into water.

Evaluation of results and questions to consider

- If a scan of wavelengths was performed, was a distinct peak at 280 nm evident?
- Calculate the molar concentration of the siderophore in the preparation prior to dilution for spectrophotometry. If your technique was good, millimolar will probably be appropriate. The following should be used.
  1. The A_{280} of the diluted siderophore.
  2. The ε_{280} of aerobactin that was stated earlier in this exercise.
  3. The path length in the cuvette.
  4. The Beer−Lambert law as described in the earlier chapter that provided an introduction to the use of UV radiation for spectrophotometry.
  5. The dilution ratio for the sample in the cuvette.
- Using the total volume of the preparation of the siderophore, and the concentration you have just calculated, calculate the yield of siderophore in units of moles. If your technique was good, micromoles will probably be the most appropriate.
- We do not know the molar mass of this siderophore; however, an estimate of the yield in units of mass may be calculated by using the molar mass of desferric aerobactin, 564.24 g mol^{-1} (Haygood et al., 1993).

10.6 COLORIMETRIC AND SPECTROPHOTOMETRIC ASSAY FOR CHELATION OF IRON(III) BY THE SIDEROPHORE

Theoretical background

An aqueous solution of iron(III) chloride, that is, $FeCl_3$ (aq), appears pale yellow because the ions of iron(III) absorb visible radiation. After aerobactin or another siderophore is added to such a solution and chelation of iron(III) occurs, the solution appears orange, red, brown or blue (Gibson & Magrath, 1969; Küpper et al., 2006; Pecoraro et al., 1983; Young & Gibson, 1979).

The molar concentration of aerobactin in a sample may be quantified by mixing an aliquot with an aliquot of $FeCl_3$ (aq) and assaying the A_{500}. The ε_{500} for ferric aerobactin has been reported to be $2.0\ mM^{-1}\ cm^{-1}$ (Gibson & Magrath, 1969).

Technical background

Assaying for the interaction of two molecules in solution by observing a change in color is known as *colorimetry*. In this case the *colorimetric assay*, as described above, will be positive if chelation of iron(III) occurs. This may be observed with the unaided eye. Samples of just a few microliters are sufficient for this assay; however, for spectrophotometric quantification of the expected increase in the A_{500}, the mixture of siderophore and $FeCl_3$ (aq) must be diluted to a volume appropriate for the cuvette. This issue is explained in the earlier exercise in this chapter involving spectrophotometric quantification with UV radiation.

The molar ratio of iron(III) to siderophore in these assays will be 6:1. To verify that the new color and the A_{500} of the mixture of the siderophore and $FeCl_3$ (aq) is due to absorption by the ferric siderophore rather than the desferric siderophore or unchelated ions of iron(III), you should assay samples of each substance by itself in addition to the mixture.

If a spectrophotometer capable of recording a scan of the visible range of wavelengths (400–700 nm) is available, such a scan should be recorded for all three samples. This will verify that 500 nm is the best wavelength for quantifying absorption of the ferric siderophore. If your spectrophotometer is only capable of recording absorbance at a fixed wavelength, assaying the A_{500} will be sufficient.

Materials

Reagents

- An aliquot of the siderophore preparation that was eluted from the solid phase and then further purified by precipitation of undesired substances. The molar concentration of the siderophore should have been quantified in an earlier exercise. Two aliquots with 4.0 nmol in each should be sufficient.
- 12.0 mM $FeCl_3$ (aq) (corresponds to 12.0 nmol μL^{-1} or 2.0 $\mu g\ \mu L^{-1}$). Prepare a convenient volume. If good data are obtained on the first try, only 4.0 μL will be needed. Prepare within 24 h of the time it will be used. Minimize exposure to atmospheric oxygen.

Supplies and equipment

- A spectrophotometer capable of quantifying absorbance of visible radiation. The most critical wavelength will be 500 nm; however, a device that will record a spectrum from 400 to 700 nm and display it by means of associated software is preferred.
- Miniature test tubes or a multiwell plate, made of transparent plastic, for the observation of the colorimetric assay with the unaided eye.
- One or two cuvettes as described in the earlier exercise in this chapter in which spectrophotometry with UV radiation is performed. For visible radiation glass cuvettes may be sufficiently transparent; however, if quartz cuvettes are available they should be used.

Procedure

Prepare the samples for colorimetry and spectrophotometry and collect data (Table 10.2).

Evaluation of results and questions to consider

- Did the colorimetric assay provide evidence for chelation?
- If a scan of the entire visible region of radiation was performed, does 500 nm appear to be the best wavelength for determining if the mixture of the siderophore and $FeCl_3$ (aq) absorbs visible radiation more strongly than solutions containing either substance alone?

TABLE 10.2 Colorimetric and spectrophotometric assay for the chelation of iron(III) by the siderophore from _E. aerogenes_.

	Siderophore	FeCl₃ (aq)	Siderophore and FeCl₃ (aq)
_____ mM siderophore from _E. aerogenes_, in methanol (4.0 nmol per sample)	_____ μL	None	_____ μL
12.0 mM FeCl₃ (aq) (24.0 nmol per sample)	None	2.0 μL	2.0 μL
Methanol (equal to volume of siderophore in the other two samples)	None	_____ μL	None
	Observe and record color of each sample		
Distilled water	_____ μL	_____ μL	_____ μL
Total volume (minimum for spectrophotometry in cuvettes, should be equal for all three samples)	_____ μL	_____ μL	_____ μL
	Scan from A_{400} to A_{700}, or quantify A_{500}, for each sample		

- Compare the A_{500} of each of the three samples. Does this spectrophotometric assay suggest that chelation has occurred? Explain your reasoning.

10.7 CHROMATOGRAPHIC ASSAY FOR CHELATION OF IRON(III) BY THE SIDEROPHORE

Theoretical background

If the substance that has been purified is aerobactin, or a siderophore with a similar structure, it should be a linear molecule that contains two or more hydroxamic acid groups. The time of retention of such a siderophore in a high-performance liquid chromatography (HPLC) experiment has been shown to be decreased by chelation of iron(III) (D'Onofrio et al., 2010). There are probably two reasons for the decrease in the time of retention. The first is that the ferric form of the siderophore is more polar than the desferric form. The second is that the ionic bonds formed between the hydroxamic acid groups and the iron(III) in the ferric form cause the siderophore to wrap around the metallic ion, thus making the siderophore more compact. The stationary phase in the chromatographic column used in HPLC consists of nonpolar chains of hydrocarbon. The more polar and more compact ferric siderophore does not associate as tightly with the stationary phase as does the desferric form. Comparison of a chromatographic assay of a mixture of a siderophore and aqueous $FeCl_3$ with that of the siderophore by itself may provide evidence for chelation.

Technical background

An earlier chapter in this book introduces HPLC. As explained there, the most modern version of this technique is ultra-high-performance liquid chromatography (UHPLC). In that earlier chapter an example of data that was obtained for a siderophore from *E. coli* was provided. The parameters for the chromatographic process were also provided there. The procedure provided in this exercise for examination of the siderophore from *E. aerogenes* is written with the assumption that a chromatographic device similar to the one described in the earlier chapter mentioned above will be used. If the device to be used is significantly different, appropriate modifications should be made.

Also mentioned in the earlier chapter in this book that was referred to above, is the inclusion of acid in the mobile phase in this type of chromatography. The observation that siderophores bind iron(III) more efficiently in a solution at a pH near neutral rather than one with a low pH (Pecoraro et al., 1983; Young & Gibson, 1979) suggests that it is best to not include acid in the mobile phase if assaying for chelation. In the procedure provided below, the mobile phase does not contain acid.

As explained in another exercise in this chapter, the desferric form of the siderophore that has been purified should absorb UV radiation with a wavelength of 280 nm. Iron(III) that is released when solid $FeCl_3$ is dissolved in water, and the ferric form of the siderophore, also absorb this wavelength. To quantify the retention time of the desferric and ferric forms of the siderophore, the A_{280} of the analytes exiting the column should be recorded.

To prepare the samples of analyte for this experiment it is essential to have an estimate of the concentration in both mass per volume and molarity. In the exercise presented earlier in this book in which spectrophotometry was used to quantify the concentration of the desferric siderophore, an extinction coefficient, ε_{280}, was provided to allow for calculation of the molarity. As is explained in earlier exercises in this chapter, the structure of the siderophore that is secreted by *E. aerogenes* is not yet known; however, it may be similar to aerobactin. To convert the value of the concentration between molarity and mass per volume, the molar mass of desferric aerobactin, $564.24\,g\,mol^{-1}$ (Haygood et al., 1993), may be used as an estimate of the molar mass of the desferric siderophore from *E. aerogenes*.

Materials

Reagents
- An aliquot of the siderophore that was eluted from the solid phase in methanol and then clarified by precipitation with ammonium hydroxide. A sample of 6.0 µg is needed.
- An aqueous solution of 6.2 mM ($1.0\,mg\,mL^{-1}$) $FeCl_3$. Prepare at least 120.0 µL, on the day the experiment is to be performed.
- A small volume, no more than 1.0 mL, of methanol. A small aliquot of this will be needed to prepare the sample of aqueous $FeCl_3$ with no siderophore as shown in Table 10.3.
- Solvents for the mobile phase of the chromatographic process.

  Solution A: deionized water, pH between 5.0 and 7.0.
  Solution B: acetonitrile, pH ~7.0.

Supplies and equipment
- Micropipettors and plastic pipette tips.
- *Autosampler vials* for loading samples of analyte into the chromatographic device. These vials are constructed of plastic or glass. The internal compartment may have a cylindrical or conical shape. The conical shape is preferable if there is a concern that the volume of the sample may be too small for proper uptake into the needle that will subsequently inject the sample into the column.
- An HPLC or UHPLC device connected to a computer in which an application for operation of the device and collection of data has been installed.

TABLE 10.3 Chromatographic assay for the chelation of iron(III) by the siderophore from *Enterobacter aerogenes*.

	Siderophore	FeCl$_3$ (aq)	Siderophore and FeCl$_3$ (aq)
_____ mM Siderophore from *E. aerogenes*, in methanol (3.0 μg per sample)	_____ μL	None	_____ μL
6.2 mM FeCl$_3$ (aq) (60.0 μg per sample)	None	60.0 μL	60.0 μL
Methanol (equal to volume of siderophore in the other two samples)	None	_____ μL	None
	Allow 5.0 min for chelation to occur		
Distilled water	_____ μL	_____ μL	_____ μL
Total volume	200.0 μL	200.0 μL	200.0 μL
Volume to inject into chromatographic device	10.0 μL	10.0 μL	10.0 μL
Mass of siderophore injected	0.150 μg	None	0.150 μg
Mass of FeCl$_3$ injected	None	3.0 μg	3.0 μg
Moles of siderophore injected	0.266 nmol	None	0.266 nmol
Moles of iron(III) injected	None	18.5 nmol	18.5 nmol
Molar ratio of siderophore: iron(III)	Not applicable	Not applicable	1:70

Procedure

- Follow the instructions provided in Table 10.3 to prepare the samples and inject appropriate portions into the chromatographic device. The three samples should be: the siderophore by itself, aqueous FeCl$_3$, and a mixture of the siderophore and aqueous FeCl$_3$.
- In the chapter that introduced HPLC and UHPLC that was presented earlier in this book, the characteristics of a typical chromatographic column used in this technique are described. Two protocols for the process of chromatography

are presented, one for a 3-min run and other for a 10-min run. If a column that is similar to the one described earlier is used, the protocol for a 3-min run will be appropriate for this experiment. As explained earlier in *Technical background*, the mobile phase should *not* contain trifluoroacetic acid and the A_{280} of substances eluting from the column should be monitored.

Evaluation of results and questions to consider

- Does the time of retention of the siderophore that was injected into the column without prior addition of aqueous $FeCl_3$ appear to be different than the time of retention of iron (III) that has not been chelated? If there is a difference, what is the most likely explanation?
- Is the time of retention of the siderophore that was mixed with aqueous $FeCl_3$ prior to injection different than that of the siderophore with no $FeCl_3$?
- Is the data that were obtained in this experiment consistent with the conclusion that the substance that was secreted by *E. aerogenes* and purified in earlier exercises in this book is a siderophore? Explain your reasoning.

10.8 ASSAY BY MASS SPECTROMETRY FOR CHELATION OF IRON(III) BY THE SIDEROPHORE

Theoretical background

A chapter presented earlier in this book provided an introduction to mass spectrometry (MS). As explained in that chapter, a molecule must be in an ionic form to be projected though the electric field toward the detector. The uppercase letter "M" is used to represent a molecule of neutral analyte (Niessen, 2006).

For MS, there are several ways in which a neutral organic molecule may be converted to an ion. Two of these ways are the addition to, or loss of a proton from, this molecule. This generates a *protonated molecule*, indicated by the notation +H, or a *deprotonated molecule* indicated by the notation −H. The net charge on the ionic form is denoted by a plus or minus sign in superscript. The protonated form is thus denoted $[M+H]^+$, whereas $[M−H]^-$ represents the deprotonated form. To detect $[M+H]^+$ the mass spectrum is recorded in *positive-ion mode* whereas *negative-ion mode* is used to detect $[M−H]^-$. Both desferric and ferric aerobactin have been detected by MS in positive-ion mode (Küpper et al., 2006). MS study of other hydroxamate-containing siderophores has also been performed in positive-ion mode (D'Onofrio et al., 2010).

Technical background

For a pure substance, good data may be obtained by injecting it directly into an MS device. If the analyte is expected to be a mixture, or if it is possible that the purification method did not remove all of the undesired substances, the hybrid technique of liquid chromatography–mass spectrometry (LC–MS) should be used. If the chromatography parameters are set properly, the LC will separate the substance of interest from other substances. In this way a mass spectrum of the pure substance of interest may be obtained.

As explained in the earlier chapter that was mentioned above, the data produced by MS is a *mass spectrum*. The symbol m represents the molar mass of an analyte in grams per mole and the symbol z represents the number of charges on each molecule of this analyte. A mass spectrum is a histogram of the relative amount of analyte molecules that have a particular m/z ratio.

As explained in the chapter presented earlier in this book in which HPLC was introduced, acid is typically included in the mobile phase for this type of chromatography. The observation that siderophores bind iron(III) more efficiently in a solution at a pH near neutral rather than one with a low pH (Pecoraro et al., 1983; Young & Gibson, 1979) suggests that it is best to not include acid in the mobile phase if assaying for chelation. In the procedure provided below, the mobile phase does not contain acid.

LC–MS produces two types of data. For LC, a *chromatogram* in which the amount of analyte eluting from the column is plotted as a function of time is generated. For analytes that efficiently absorb UV or visible radiation, spectrophotometry may be used to generate the chromatogram. Some of these devices also have a *mass detection* feature that is not dependent on absorption of radiation by the analyte. In previous exercises in this book the A_{280} was used to detect both the ferric and desferric forms of the siderophore from *E. aerogenes*. The A_{500} was used to specifically detect the ferric form. If the LC component is included in the current exercise, A_{280} or mass detection may be used to detect either the ferric or desferric form of the siderophore. A_{500} may only be used to detect the ferric form.

If there is single peak on the chromatogram, it should include the siderophore. A mass spectrum of the substance in this peak should be examined. If there are multiple peaks, the mass spectrum of the peak that is believed to be the siderophore should be examined.

The siderophore that has been purified is dissolved in methanol. As shown in the table presented below in *Procedure*, when aliquots of this sample are prepared for MS or LC−MS, they are diluted into deionized water. As explained below in *Materials*, detection of organic metabolites with MS is very sensitive. The dilution will probably be at least 50-fold and thus the concentration of methanol in the diluted solution will probably be less than 2.0%. It is unlikely that this amount of the organic solvent will interfere with either the chromatography or the spectrometry.

The structure and molar mass of the siderophore from *E. aerogenes* are not yet known. To convert the value of the concentration of the siderophore between molarity and mass per volume, the molar mass of aerobactin, $564.24 \, g \, mol^{-1}$ (Haygood et al., 1993) may be used. The justification for this is provided in *Technical background* in the exercise presented earlier in this book in which chelation was assayed with HPLC.

Materials

Reagents

- An aliquot of the siderophore that was eluted from the solid phase in methanol and then clarified by precipitation with ammonium hydroxide. Because the amount of analyte necessary to obtain good data varies depending on the type of device employed and whether the MS or LC−MS technique is performed, your instructor and the laboratory staff will have to provide advice. The table presented in *Procedure* specifies amounts of analyte that may be suitable.

A typical LC−MS assay of a bacterial secondary metabolite requires only 150.0−300.0 pmol of analyte. Using the molar mass of aerobactin cited earlier in *Technical background*, this number of picomoles corresponds to between 85.0 and 170.0 ng. If samples of analyte are prepared as specified in the table presented in *Procedure*, 6.0 µg of the siderophore will be needed. The aliquots injected will each contain 150.0 ng of the siderophore.

- An aqueous solution of 6.2 mM (1.0 mg mL^{-1}) FeCl$_3$. Prepare at least 60.0 µL, on the day the experiment is to be performed.
- Solvents for the mobile phase of the chromatographic process (if LC−MS will be performed).

  Solution A: deionized water, pH between 5.0 and 7.0.
  Solution B: acetonitrile, pH ~7.0.

Supplies and equipment

- Micropipettors and plastic pipette tips.
- *Autosampler vials* for loading samples of analyte into the MS or LC−MS device. These vials are constructed of plastic or glass. The internal compartment may have a cylindrical or conical shape. The conical shape is preferable if there is a concern that the volume of the sample may be too small for proper uptake into the needle that will subsequently inject the sample.
- An MS or LC−MS device connected to a computer in which an application for operation of the device and collection of data has been installed.

Procedure

- Follow the instructions provided in Table 10.4 to prepare the samples for injection into the MS or LC−MS device. The two samples should be: the siderophore by itself and a mixture of the siderophore and aqueous FeCl$_3$.

Operation of the analytical device

- If MS will be performed.
  Follow the instructions provided by the instructor, laboratory staff and the manufacturer of the device.
- If LC−MS will be performed.

In the chapter that introduced HPLC and UHPLC that was presented earlier in this book, the characteristics of a typical chromatographic column used in this technique are described. Two protocols for the process of chromatography are presented, one for a 3-min run and other for a 10-min run. If a column that is similar to the one described in that earlier chapter is used, the protocol for a 3-min run will be appropriate for this experiment. As explained earlier in *Technical background*, the mobile phase should not contain trifluoroacetic acid and there are three methods by which the siderophore may be detected as it elutes from the column.

TABLE 10.4 Assay by mass spectrometry for the chelation of iron (III) by the siderophore from *Enterobacter aerogenes*.

	Siderophore	Siderophore and FeCl$_3$ (aq)
_____ mM Siderophore from E. aerogenes, in methanol (3.0 µg per sample)	_____ µL	_____ µL
6.2 mM FeCl$_3$ (aq) (60.0 µg per sample)	None	_____ µL
	Allow 5.0 min for chelation to occur	
Distilled water	_____ µL	_____ µL
Total volume	200.0 µL	200.0 µL
Volume to inject into device for mass spectrometry or liquid chromatography	10.0 µL	10.0 µL
Mass of siderophore injected	0.150 µg	0.150 µg
Mass of FeCl$_3$ injected	None	3.0 µg
Moles of siderophore injected	0.266 nmol	0.266 nmol
Moles of iron(III) injected	None	18.5 nmol
Molar ratio of siderophore:iron(III)	Not applicable	1:70

Collection of data

- MS.
     A mass spectrum should be recorded.
- LC–MS.

At least one chromatogram and at least one mass spectrum should be recorded. If multiple detection methods are used for LC, multiple chromatograms should be obtained. If multiple peaks that have eluted from the column are examined with MS, multiple mass spectra should be obtained.

Evaluation of results and questions to consider

- If the *m/z* ratios observed suggest that the siderophore that has been purified is aerobactin, the following should be done.

Refer to the articles that report the assay of ferric and desferric aerobactin with MS. The most recent of these (Küpper et al., 2006) is the most useful (the spectra summarized in this article are available without a fee as supplementary information on the publisher's website). Articles that were published earlier may also be helpful (Buyer, de Lorenzo, & Neilands, 1991; Haygood et al., 1993). Discuss whether or not your data are similar to the published results for aerobactin. Discuss whether or not your results show that the siderophore that has been purified chelates iron(III).

- If the m/z ratios observed suggest that the siderophore that has been purified is *not* aerobactin, the following should be done.

  Discuss whether or not the m/z ratios provide evidence that the organic molecule that has been purified chelates iron(III).

  Refer to: the articles cited in the bibliographies of this book, other articles that discuss siderophores and chemical databases. One of the articles (D'Onofrio et al., 2010) cited here discusses the structures of many siderophores that include hydroxamic acid. Molar masses for the ferric and desferric forms of five of these siderophores, quantified by MS, are summarized in the supplementary information. Discuss whether or not the observed m/z ratios suggest a molar mass that is in the range typical of siderophores.

10.9 INFRARED SPECTROSCOPIC ELUCIDATION OF THE STRUCTURE OF THE SIDEROPHORE

Theoretical background

In earlier exercises in this chapter the siderophore is initially enriched in a low pH (~ 2.3) solution and then undesired substances are removed by precipitation with ammonium hydroxide. The precipitation also increases the pH of the solution to approximately 8.0. Although the higher pH improves the chelation of iron(III) by this siderophore, it is not ideal for IR spectroscopy.

The spectrum of absorption of IR radiation by aerobactin has been published (Gibson & Magrath, 1969). In this publication the pH of the solution in which aerobactin was dissolved prior to the assay was not specified. I have observed that the IR spectrum of the siderophore from *E. aerogenes* is affected by pH.

These assays were performed on dry samples; however, the pH of the solution containing the siderophore was measured before evaporation of solvent. The IR spectrum obtained for siderophore from a solution at pH 2.0 resembles more closely the published spectrum of aerobactin than does the spectrum obtained from a solution at pH 8.0. The siderophore purified by the method in this chapter may be aerobactin or have a structure similar to aerobactin; therefore, it is best if the sample to be assayed has a pH of 2.0.

Materials

- An aliquot containing 4.0 nmol of the siderophore (in methanol) will be needed for each assay with the IR spectrometer. If the molar mass of the siderophore that has been purified is close to that of aerobactin ($564.24 \, g \, mol^{-1}$) (Haygood et al., 1993), this aliquot will contain $\sim 2.3 \, \mu g$. Be certain that the amount of this sample necessary for the number of assays that will be performed is available for use.
- An aqueous solution of 5.0 M hydrochloric acid.

Procedure

- To prepare the sample for spectroscopy, an appropriate volume should be acidified by addition of 5.0 M HCl (aq). Add a volume of acid that is one-ninth that of the solution of the siderophore and mix. The final concentration of HCl should be 0.5 M and the pH should be close to 2.0. The final pH may be confirmed by assaying a drop of the siderophore solution on pH indicator paper.
- If the device being used is designed to assay dry samples, place a drop containing 4.0 nmol of the siderophore (in methanol) onto the sample-loading spot. Allow the solvent to evaporate.
- Acquire the spectrum of absorption of IR radiation.

Evaluation of results and questions to consider

- Compare the IR spectrum that you have recorded to the published spectrum for aerobactin (Gibson & Magrath, 1969) (panel B in Figure 2 in this article). Is it possible that aerobactin or a siderophore with a similar structure has been purified?

10.10   NUCLEAR MAGNETIC RESONANCE SPECTROSCOPIC ELUCIDATION OF THE STRUCTURE OF THE SIDEROPHORE

Theoretical background

As explained in the chapter presented earlier in this book in which this technique was introduced, there are several isotopes to which NMR spectroscopy may be applied. In organic substances the most abundant of these is the predominant isotope of hydrogen, ^1H. In this exercise ^1H NMR spectroscopy of the analyte, the siderophore that was purified in earlier exercises, is performed. This technique is also known as *proton NMR*.

A molecule or atom that has one or more unpaired electrons is *paramagnetic* (Atkins, Jones, & Laverman, 2016; Keeler, 2010). Substances comprised of atoms in which all of the electrons are paired are *diamagnetic*. A paramagnetic substance is pulled into a magnetic field whereas a diamagnetic substance is repelled by a magnetic field. Molecular oxygen, O_2, and iron (III) both have at least one unpaired electron and therefore expected to be paramagnetic.

When a pulse of radio-frequency radiation from the excitation coil in the transmitter is applied to the analyte during NMR spectroscopy, some of the energy in this radiation is absorbed by atomic nuclei (Keeler, 2010; Richards & Hollerton, 2011). This absorption can cause some of the nuclei to enter an *excited spin state* in which they have a higher potential energy than they did in the *ground spin state* prior to application of the pulse. After the pulse has ended, these nuclei return to the ground state by the process of *relaxation*.

The presence of a paramagnetic atom or molecule in a sample of analyte being examined with NMR spectroscopy causes relaxation to occur more rapidly than it usually does and this decreases the resolution of the spectrum (Field, Sternhell, & Kalman, 2013; Keeler, 2010; Richards & Hollerton, 2011). When using NMR to examine the structure of a siderophore it is best to apply this technique to the desferric form to ensure that high quality data will be obtained.

Technical background

Refer to the chapter presented earlier in this book in which NMR spectroscopy was introduced and to some of the cited references (Richards & Hollerton, 2011) for a discussion of appropriate solvents. One of the goals in choosing a solvent is minimization of resonance peaks in the spectrum due to the molecules of solvent. For ^1H NMR spectroscopy the analyte, a siderophore in this case, should be dissolved in a solvent in which the atoms of ^1H have been replaced with ^2H (deuterium). This is known as a *deuterated solvent* and it is used because ^2H does not generate a peak of resonance on the spectrum in this type of experiment. Deuterated versions of water or methanol are often used as solvents for organic metabolites being examined with NMR spectroscopy. The disadvantage of these two solvents is that exchange between the ^1H in acidic groups in the analyte and the ^2H in the solvent results in loss of the expected peaks of resonance for these protons. In the siderophore that was purified in earlier exercises in this book, the acidic groups that are likely to be present are hydroxamic acid, carboxylic acid, hydroxyl and amine.

The solvents most commonly used for NMR spectroscopy of siderophores that include hydroxamic acid are deutero-water (D_2O) and deutero-dimethyl sulfoxide (D_6-DMSO) (Borgias, Hugi, & Raymond, 1989; Buyer et al., 1991; D'Onofrio et al., 2010; Gibson & Magrath, 1969; Harris et al., 1979; Küpper et al., 2006). Because of the exchange of ^1H from acidic groups in molecules of analyte with ^2H in D_2O that was mentioned earlier, D_6-DMSO is the preferred solvent for the siderophore to be examined in this experiment.

Peaks of resonance due to the solvent are not completely eliminated by the use of ^2H. As explained in the earlier chapter mentioned above, some exchange of ^1H in the analyte with ^2H in the molecules of solvent will result in a solvent-specific peak or cluster of peaks in the NMR spectrum and this may complicate interpretation of the data. In an experiment in which an analyte is dissolved in D_6-DMSO be aware that under atmospheric pressure the boiling point of this liquid is 189.0°C After a solution in D_6-DMSO is prepared it is difficult to increase the concentration of the solute because this solvent does not readily evaporate.

Materials

Reagents

- D_6-DMSO to dissolve the analyte, that is, the siderophore that has been purified, as explained in *Technical background* earlier in this exercise. As discussed below, each assay usually requires less than 1.0 mL of solvent.

• An aliquot of the desferric siderophore that was eluted from the solid phase in methanol and then clarified by precipitation with ammonium hydroxide. If the resonance frequency of the NMR spectrometer is 500.0 MHz or greater, 6.0 μmol will be sufficient. If the molar mass of aerobactin, 564.24 g mol^{-1} (Haygood et al., 1993) is used as an estimate of the molar mass of the siderophore that has been purified, 3.4 mg will correspond to 6.0 μmol. If the siderophore is resuspended in D_6-DMSO at a concentration of 10.0 mM then 0.6 mL will contain 6.0 μmol. This is a typical volume for an NMR spectroscopic experiment.

Supplies and equipment

• Micropipettors and pipette tips.
• A test tube for evaporation of the methanol solvent from the sample of analyte and resuspension in the D_6-DMSO.
• A fume hood or a centrifuge in a low-pressure chamber for evaporation of the methanol.
• An *NMR sample tube*, that is, cylindrical glass tube designed to hold a solution of analyte that will be assayed with NMR spectroscopy. A tube with an outside diameter of 5.0 mm and a length of 18.0 cm is appropriate for a sample of 0.6 mL.
• An NMR spectrometer.

Procedure

• Consult with the instructor and/or the member of the laboratory staff that maintains the NMR spectrometer regarding the amount of analyte that will be necessary to obtain a good spectrum. Determine the number of moles necessary and the appropriate volume for the sample.
• Measure the appropriate volume of the solution of desferric siderophore (in methanol) into a tube that will allow for efficient evaporation of solvent.
• Leave the tube uncovered in a fume hood or use a centrifuge in an low-pressure chamber to remove the solvent by means of evaporation.
• Resuspend the analyte in the D_6-DMSO and then use a micropipettor to transfer the sample into the NMR tube.
• Load the NMR tube that contains the analyte into the spectrometer and collect the data.

Evaluation of results and questions to consider

Note: When evaluating NMR spectra of siderophores in the scientific literature, be aware that in some cases the analyte was dissolved in D_2O. As explained earlier in *Technical background*, if the analyte is dissolved in this solvent no peaks of resonance will be detected for ^1H on the acidic groups such as hydroxamic acid, carboxylic acid, hydroxyl and amine.

• Compare the spectrum that has been recorded to the NMR spectra that have been published for aerobactin (Buyer et al., 1991; Gibson & Magrath, 1969; Haygood et al., 1993; Küpper et al., 2006). Discuss whether or not this NMR spectrum is evidence that the siderophore that has been purified may be aerobactin.

• If the spectrum that has been recorded suggests that the siderophore that has been purified is *not* aerobactin, compare it to the spectra that have been published for other siderophores that contain hydroxamic acid (Borgias et al., 1989; D'Onofrio et al., 2010). (In the article published in 2010, the NMR spectra are in the supplementary information.) Discuss whether or not this NMR spectrum is evidence that the siderophore that has been purified may correspond to one of those that has been previously characterized.

10.11 GROWING CRYSTALS OF FERRIOXAMINE E, A COMPLEX OF A SIDEROPHORE AND IRON(III) FROM THE BACTERIUM *STREPTOMYCES ANTIBIOTICUS*

Experimental plan and schedule of activities

The protocol provided below is known to produce crystals of ferrioxamine E that are suitable for study by X-ray diffraction. Your instructor may have some students in your class perform these steps exactly as written but direct others to use variations of this procedure. With such an approach the class may create a procedure that generates a greater yield of crystals or is simpler than the procedure that is provided.

Using the protocol written below, four weeks will be required to complete this exercise; however, most of this time involves waiting for solvent to evaporate and crystals to form. You will probably be instructed to perform other exercises during the waiting periods to make good use of scheduled laboratory time.

Theoretical background

Ferrioxamine E is an organometallic compound consisting of a siderophore bound to a single ion of iron(III). The organic portion is the *ligand*. In this and subsequent exercises, ferrioxamine E will sometimes be referred to as the *analyte*. The molecular formula and molar mass of the compound may be obtained from the scientific literature or a chemical database.

Technical background

General considerations in the process of growing crystals

Growth of crystals is favored by minimizing disturbances of the solution of analyte (Laudise, 1970; Massa, 2004; Stout & Jensen, 1989). The preparation should be handled as gently as possible. To add solvent, gently pipette the liquid onto the surface of the solution of analyte to form a new layer. If the new solvent is miscible with the solvent in which the analyte is dissolved, the two layers will gradually mix by diffusion.

The properties of ferrioxamine E

It is soluble up to 1.0 mg mL^{-1} in water (Sigma-Aldrich_Co., 2017).

Materials

Reagents

- Ferrioxamine E. A mass of 2.0 mg of granular solid in a glass vial that has a flat bottom and screw cap (20−25 mm in diameter, 40−60 mm in height).
- Methanol. A volume of approximately 20.0 mL that is pure enough for HPLC.
- The most pure water that is available should be used. The ideal is water that has been purified twice by distillation; however, if the best water available is that which has been distilled once or that which has been purified by filtration, it may be used.

Supplies

- If the vial containing the ferrioxamine E does not have graduations for measuring the volume of liquid, an empty vial of the same dimensions should be obtained.

Procedure

- If the vial containing the ferrioxamine E is not graduated for measuring the volume of enclosed liquid, create graduations at intervals of 2.0 mL with a felt-tip pen. Use the empty vial and known volumes of water as a guide.
- Day 1.

  Dissolve the 2.0 mg of ferrioxamine E in 2.0 mL of 5% (v/v) methanol. Leave the vial open to the atmosphere, in a location where it is shielded from direct light and will not be disturbed. During the subsequent days, evaporation of solvent and increase of the concentration of analyte should occur.
- Day 10.

  Using the graduations on the vial, and without disturbing the solution of analyte, estimate and record the remaining volume of liquid. Also record the shape, size, and color of any precipitate that has formed.

  Gently layer 2.0 mL of pure methanol onto the surface of the solution of analyte. Methanol is miscible with water and gentle application of the methanol will result in a gradual homogenization of the two solvents.

  Cover the vial with the screw cap and, as before, leave and it in a protected location. Allow approximately 24 h for gradual mixing of solvents.
- Day 11.

  Because methanol is more volatile than water, the small percentage of methanol in the original solution of analyte probably evaporated between day 1 and day 10. Calculate the concentration of methanol (volume to volume percentage) that is now in the solution. Record the shape, size, and color of any precipitate that has formed.

  Uncover the vial and gently layer 4.0 mL of pure methanol onto the surface of the solution of analyte. Record the final volume of liquid in the vial.

  Place the cap back on the vial and leave it in a protected location. Allow approximately 24 h for gradual mixing of solvents.
- Day 12.

  Calculate the concentration of methanol that is now in the solution. Record the shape, size, and color of any precipitate that has formed.

  Uncover the vial and leave it in a protected location to allow for evaporation and increase of the concentration of analyte.

- Day 17.

  Record the volume of liquid in the vial. Assume that the methanol evaporates before the water. Use the values of the concentration of methanol and the total volume on day 12, and the total volume today, to calculate if there is any methanol remaining in the solution. Record the shape, size, and color of any precipitate that has formed.

  Calculate the volume of methanol that must be added to the solution of analyte so that after the solution becomes homogeneous the concentration of methanol will be 80% (v/v). Gently layer this volume of methanol onto the solution of analyte. Record the final volume of liquid in the vial. Place the cap back on the vial and store it in a protected location for approximately 24 h to allow the solvents to gradually mix.

- Day 18.

  The solution of analyte should again be homogeneous. Remove the cap from the vial and leave it in a protected location to allow for evaporation and increase of the concentration of analyte.

- Day 28.

  Record the volume of liquid in the vial. Is it likely that there is any methanol remaining in the solution? Hopefully a precipitate has formed in the preparation. If so, record the shape, size, and color.

  Do the following to store the precipitated solid until you are ready to collect X-ray diffraction data.

- Add enough methanol to the vial to bring the final concentration to 95%.
- Secure the cap on the vial to prevent evaporation.
- Store the vial at ambient laboratory temperature, shielded from light.

Evaluation of results and questions to consider

- Does the initial solution of ferrioxamine E show a color? If so, which component of the complex generates the color?
- If the result of mixing equal volumes of two solvents is a homogeneous solution, that is, no interface separating two layers, the solvents are said to be miscible. What would be the effect of substituting butanol for methanol in this experiment?

10.12 EVALUATION OF THE PRECIPITATED FERRIOXAMINE E FOR THE PROPERTIES OF CRYSTALS

Theoretical background

Refraction is the alteration of the path of waves of light that occurs when the light passes between media of different densities. This phenomenon can occur when light passes through a transparent solid regardless of whether or not it is a crystal. For example, silicon dioxide (SiO_2) is a liquid at very high temperature, but when allowed to cool in a way that the molecules of SiO_2 form covalent bonds with one another in a random array, glass is formed. Glass is not a crystal but rather an *amorphous* solid because the arrangement of the molecules of SiO_2 is random. Glass is more dense than air. Light that passes through air and then through glass will be refracted if the path of the light is oblique (i.e., not perpendicular) to the surface of the glass. Refraction also occurs when light passes through crystals of sodium chloride. The Na^+ and Cl^- atoms are arranged in a cubic lattice making this an *isotropic* crystal.

Birefringence, also known as *double refraction*, occurs when light passes through media of different densities and one of the media is an *anisotropic* crystal (Laudise, 1970; Massa, 2004; Nikon_Instruments_Inc., 2017; Stout & Jensen, 1989). In such a crystal the atoms may be arranged in one of several lattices, but not in a cubic lattice. If the wave of light entering the crystal (incident light) is not polarized, it will become polarized within the crystal. It will also be split into two paths, the *ordinary ray* and the *extraordinary ray*. As they pass through and then exit the crystal, the planes of polarization of these two rays are perpendicular to one another. If the incident light is polarized, transit through the crystal will generate ordinary and extraordinary rays as above, and both of these polarized rays will be in an orientation distinct from the incident light.

Technical background

Birefringence is best assayed with an optical microscope designed for polarized light (Laudise, 1970; Massa, 2004; Nikon_Instruments_Inc., 2017; Stout & Jensen, 1989). Such a device has a polarizer between the source of light and the specimen, and an analyzer between the specimen and the observer's eyes. With such a device, a single plane of light enters the sample from below. The analyzer may be rotated to specify the orientation of the plane of light emitted from the specimen that will be allowed to pass through to the observer's eyes.

If the specimen is an anisotropic crystal, the emitted light will be polarized into two mutually perpendicular planes. As one complete rotation (360 degrees) of the analyzer is made, there will be four positions at which the intensity of light passing through to the observer's eyes is maximum. These four positions are at intervals of 90 degrees.

Materials

Reagents

- The solid ferrioxamine E that precipitated in an earlier exercise.

Equipment

- An optical microscope designed for polarized light.

Procedure

- Examine the solid ferrioxamine E for birefringence using polarized light microscopy.

Evaluation of results and questions to consider

- Discuss whether or not the solid ferrioxamine E appears to be a crystal.
- Discuss with your instructor whether or not it will be worthwhile to obtain X-ray diffraction data for this substance.

10.13   COLLECTION AND INTERPRETATION OF X-RAY DIFFRACTION DATA FROM SINGLE CRYSTALS OF FERRIOXAMINE E

Theoretical and technical background

- The following computational programs are commonly used for interpretation of X-ray diffraction data.

SHELX (Sheldrick, 2008)
SHELXC/D/E (Sheldrick, 2010)
SHELXL (Sheldrick, 2015a)
SHELXT (Sheldrick, 2015b)
Oak Ridge Thermal Ellipsoid Prediction (ORTEP) (Farrugia, 1997)
WinGX (Farrugia, 1999)

Evaluation of results and questions to consider

- You may need to use The International Tables for Crystallography to interpret the data obtained by X-ray diffraction of the crystal (Commission_on_International_Tables_in_the_International_Union_of_Crystallography, 2017).

BIBLIOGRAPHY

Atkins, P. W., Jones, L., & Laverman, L. (2016). *Chemical principles: The quest for insight* (7th ed.). New York, NY: W.H. Freeman: Macmillan Learning. Retrieved from <https://lccn.loc.gov/2015951706>.

Borgias, B., Hugi, A. D., & Raymond, K. N. (1989). Isomerization and solution structures of desferrioxamine B complexes of aluminum(3 +) and gallium (3 +). *Inorganic Chemistry*, *28*, 3538–3545. Available from https://doi.org/ 10.1021/ic00317a029.

Buyer, J. S., de Lorenzo, V., & Neilands, J. B. (1991). Production of the siderophore aerobactin by a halophilic pseudomonad. *Applied and Environmental Microbiology*, *57*, 2246–2250. Retrieved from <https://www.ncbi.nlm.nih.gov/pubmed/1768095>.

Commission_on_International_Tables_in_the_International_Union_of_Crystallography. (2017). *International tables for crystallography*. Chester, England. Retrieved from <http://it.iucr.org/>.

D'Onofrio, A., Crawford, J. M., Stewart, E. J., Witt, K., Gavrish, E., Epstein, S., ... Lewis, K. (2010). Siderophores from neighboring organisms promote the growth of uncultured bacteria. *Chemistry & Biology*, *17*, 254–264. Retrieved from <https://www.ncbi.nlm.nih.gov/pubmed/20338517>.

Farrugia, L. J. (1997). ORTEP-3 for Windows – A version of ORTEP-III with a graphical user interface (GUI). *Journal of Applied Crystallography*, *30*, 565. <http://onlinelibrary.wiley.com/doi/10.1107/S0021889897003117/ abstract>.

Farrugia, L. J. (1999). WinGX suite for small-molecule single-crystal crystallography. *Journal of Applied Crystallography*, *32*, 837–838. <http://onlinelibrary.wiley.com/doi/10.1107/S0021889899006020/abstract>.

Field, L. D., Sternhell, S., & Kalman, J. R. (2013). *Organic structures from spectra* (5th ed.). Chichester, West Sussex: Wiley. Retrieved from <https://lccn.loc.gov/2012046033>.

Gibson, F., & Magrath, D. I. (1969). The isolation and characterization of a hydroxamic acid (aerobactin) formed by *Aerobacter aerogenes* 62-I. *Biochimica et Biophysica Acta*, *192*, 175–184. Retrieved from <https://www.ncbi.nlm.nih.gov/pubmed/4313071>.

Harris, W. R., Carrano, C. J., & Raymond, K. N. (1979). Coordination chemistry of microbial iron transport compounds. 16. Isolation, characterization, and formation constants of ferric aerobactin. *Journal of the American Chemical Society*, *101*, 2722–2727. Available from https://doi.org/10.1021/ ja00504a038.

Haygood, M. G., Holt, P. D., & Butler, A. (1993). Aerobactin production by a planktonic marine *Vibrio* sp. *Limnology and Oceanography, 38*, 1091–1097. Available from https://doi.org/10.4319/lo.1993.38.5.1091.

Johnson, T. R., & Case, C. L. (2018). *Laboratory experiments in microbiology* (12th ed.). Hoboken: Pearson. Retrieved from <https://lccn.loc.gov/2017039734>.

Keeler, J. (2010). *Understanding NMR spectroscopy* (2nd ed.). Chichester, UK: John Wiley and Sons. Retrieved from <https://lccn.loc.gov/2009054393>.

Keller, R., Pedroso, M. Z., Ritchmann, R., & Silva, R. M. (1998). Occurrence of virulence-associated properties in *Enterobacter cloacae. Infection and Immunity, 66*, 645–649. Retrieved from <http://iai.asm.org/content/66/2/645.abstract>.

Küpper, F. C., Carrano, C. J., Kuhn, J. U., & Butler, A. (2006). Photoreactivity of iron(III)-aerobactin: Photoproduct structure and iron(III) coordination. *Inorganic Chemistry, 45*, 6028–6033. Retrieved from <https://www.ncbi.nlm.nih.gov/pubmed/16842010>.

Laudise, R. A. (1970). *The growth of single crystals*. Englewood Cliffs, NJ: Prentice-Hall. Retrieved from <https://catalog.loc.gov/vwebv/search?searchArg = Laudise&searchCode = GKEY%5E*&searchType = 0&recCount = 25&sk = en_US>.

Massa, W. (2004). *Crystal structure determination* (2nd completely updated ed.). Berlin; New York, NY: Springer. Retrieved from <http://www.loc.gov/catdir/enhancements/fy0813/2003069465-d.html>.

Niessen, W. M. A. (2006). *Liquid chromatography–mass spectrometry* (3rd ed.). Boca Raton, FL: CRC/Taylor & Francis. Retrieved from <https://lccn.loc.gov/2006013709>.

Nikon_Instruments_Inc. (2017) *Polarized light microscopy*. Melville, NY. Retrieved from <https://www.microscopyu.com/techniques/polarized-light>.

Pecoraro, V. L., Harris, W. R., Wong, G. B., Carrano, C. J., & Raymond, K. N. (1983). Coordination chemistry of microbial iron transport compounds. 23. Fourier transform infrared spectroscopy of ferric catechoylamide analogues of enterobactin. *Journal of the American Chemical Society, 105*, 4623–4633. Available from https://doi.org/10.1021/ja00352a018.

Pecoraro, V. L., Wong, G. B., Kent, T. A., & Raymond, K. N. (1983). Coordination chemistry of microbial iron transport compounds. 22. pH-dependent Moessbauer spectroscopy of ferric enterobactin and synthetic analogues. *Journal of the American Chemical Society, 105*, 4617–4623. Available from https://doi.org/10.1021/ja00352a017.

Richards, S. A., & Hollerton, J. C. (2011). *Essential practical NMR for organic chemistry*. Chichester, West Sussex: John Wiley. Retrieved from <https://lccn.loc.gov/2010033319>.

Sheldrick, G. M. (2008). A short history of SHELX. *Acta Crystallographica A, 64*, 112–122. Available from https://doi.org/10.1107/S0108767307043930.

Sheldrick, G. M. (2010). Experimental phasing with SHELXC/D/E: Combining chain tracing with density modification. *Acta Crystallographica Section D: Biological Crystallography, 66*, 479–485. Retrieved from <http://www.ncbi.nlm.nih.gov/pmc/articles/PMC2852312/>.

Sheldrick, G. M. (2015a). Crystal structure refinement with SHELXL. *Acta Crystallographica section C: Structural chemistry*, *71*, 3–8. Retrieved from <http://www.ncbi.nlm.nih.gov/pmc/articles/PMC4294323/>.

Sheldrick, G. M. (2015b). SHELXT – Integrated space-group and crystal-structure determination. *Acta Crystallographica Section A: Foundations and Advances*, *71*, 3–8. Retrieved from <http://www.ncbi.nlm.nih.gov/pmc/articles/PMC4283466/>.

Sigma-Aldrich_Co. (2017). *Ferrioxamine E, product 38266, specifications.* St Louis, MO. Retrieved from <http://www.sigmaaldrich.com/catalog/DataSheetPage.do?brandKey = SIAL&symbol = 38266>.

Stout, G. H., & Jensen, L. H. (1989). *X-ray structure determination: A practical guide* (2nd ed.). New York, NY: Wiley. Retrieved from <http://www.loc.gov/catdir/description/wiley032/88027931.html>.

Tortora, G. J., Funke, B. R., & Case, C. L. (2019). Microbiology: An introduction, (13th ed.). Boston, MA: Pearson. Retrieved from <https://lccn.loc.gov/2017044147>.

Van Tiel-Menkveld, G. J., Mentjox-Vervuurt, J. M., Oudega, B., & de Graaf, F. K. (1982). Siderophore production by *Enterobacter cloacae* and a common receptor protein for the uptake of aerobactin and cloacin DF13. *Journal of Bacteriology*, *150*, 490–497. Retrieved from <http://jb.asm.org/content/150/2/490.abstract>.

Young, I. G., & Gibson, F. (1979). Isolation of enterochelin from *Escherichia coli*. *Methods in Enzymology*, *56*, 394–398. Retrieved from <http://www.sciencedirect.com/science/article/pii/0076687979560376>.

Chapter 11

Exercises in purifying and characterizing a chloroplast-targeting phytotoxin

11.1 GROWTH OF A CULTURE OF THE GLYCINEA PATHOVAR OF THE BACTERIUM *PSEUDOMONAS SYRINGAE*

Theoretical background

The highest yield of coronatine is obtained from *Pseudomonas syringae* pathovar glycinea 4180 (Palmer & Bender, 1993). The name of this bacterial strain is often stated simply as PG4180. The media used to grow cultures of this strain are described in this exercise and the difference between *a complex medium* and a *chemically defined medium* is explained in the earlier chapter in which an overview of the methods for purification of metabolites is presented.

Technical background

Mannitol-glutamate-yeast extract (MGY) is a complex medium. PG4180 grows on MGY agar and in MGY broth. Because this medium consists of substances that are commonly present in microbiological or biochemical laboratories, it is the medium typically used to maintain stocks (agar) of PG4180 or to prepare small *starter cultures* (broth) to initiate an experiment.

The greatest yield of coronatine is obtained from a culture grown in the chemically defined medium Hoitink-Sinden optimized for production of coronatine (HSC) (Palmer & Bender, 1993). An aliquot of a starter culture that is in exponential phase is diluted into a volume of HSC broth that is much greater than the volume of the starter culture, thus creating a *subculture*. If one inoculates the large culture directly from a colony on a plate instead of using this two-step method, there is a chance that the

Purification and Characterization of Secondary Metabolites.
DOI: https://doi.org/10.1016/B978-0-12-813942-4.00011-5
147

liquid culture will not grow and the procedure will have to be repeated with another large batch of medium.

As explained in the later chapter in which a guide for design of experiments is presented, a *fluorophore* is a substance that absorbs radiation of the *excitation* wavelength and then releases radiation of the *emission* wavelength. If the emission wavelength is between 400 and 700 nm, that is, visible radiation, the substance appears to be *fluorescent* when observed with the unaided eye.

When growing in MGY medium, PG4180 bacteria secrete a fluorescent substance. When illuminated with the white lights typical in laboratories, a weak yellow-green fluorescence is visible near colonies or in the liquid culture medium. A stronger fluorescence is observed when the colonies or culture are illuminated in a dark room with UV radiation that has a wavelength of 254 nm. When this strain is grown in HSC medium, the fluorescence is not observed. This observation suggests that either the bacteria do not secrete this substance in HSC or that one of the components of HSC suppresses the fluorescence. Because the majority of bacterial species do not secrete fluorescent substances, assaying colonies or a starter culture for fluorescence is a convenient method for verifying the identity of PG4180 bacteria.

To ensure the fastest possible rate of growth of a bacterial species in the genus *Pseudomonas* in liquid culture, it is best that the microbes have access to atmospheric, molecular oxygen (O_2). The cultures should be agitated on a rotating platform or similar device to assure that oxygen is penetrating the liquid. This is specified in a procedure by stating "the culture should be incubated with aeration." If a rotating platform is used for cultures of PG4180, a frequency of between 60 and 120 rpm is appropriate.

The growth of the culture may be monitored by spectrophotometry as explained in the earlier chapter in which an overview of the methods for purification of metabolites is presented. The transmittance of visible radiation with a wavelength of 600 nm, T_{600}, or the optical density at this wavelength, OD_{600}, may be used to quantify the turbidity. This quantity may then be used to calculate the cellular density of the culture.

Materials

Reagents

- Preparation of the sterile media for the bacterial cultures is described in Tables 11.1 and 11.2.
- Colonies of the bacterial strain PG4180 on an MGY agar plate.

TABLE 11.1 Mannitol-glutamate-yeast extract (MGY) medium with agar.[a,b]

	Grams per liter	Molarity
D-Mannitol	10.0	55.0 mM
L-Glutamate	2.0	14.0 mM
K_2HPO_4	0.50	3.0 mM
$MgSO_4 \cdot 7H_2O$	0.20	0.80 mM
NaCl	0.20	3.0 mM
Yeast extract	0.25	Not applicable
Agar	15.0	Not applicable

[a]This composition was derived from the mannitol-nitrate medium that has been used for the growth of microbes in the genus Agrobacterium (Keane, Kerr, & New, 1970). The L-glutamate has been substituted for the nitrate and the concentration of yeast extract is as specified in a later publication (Palmer & Bender, 1993).
[b]Mix all components except the agar and measure the pH. It should be acidic due to the glutamate. Adjust to a pH of 7.0 with a solution of NaOH or KOH. Add the agar and sterilize in an autoclave.

TABLE 11.2 Hoitink-Sinden optimized for the production of coronatine (HSC) medium.[a]

	Mass per liter	Molarity
Glucose	20.0 g	110.0 mM
NH_4Cl	1.0 g	20.0 mM
$MgSO_4 \cdot 7H_2O$	0.20 g	0.80 mM
KH_2PO_4	4.1 g	30.0 mM
$K_2HPO_4 \cdot 3H_2O$	3.6 g	15.0 mM
KNO_3	0.30 g	3.0 mM
$FeCl_3$	3.3 mg	20.0 μM

[a]The composition of this medium has been published (Palmer & Bender, 1993) and the proper pH specified as 6.8.

Supplies and equipment

- Inoculating loop.
- Bunsen burner.
- Cuvettes for assaying the turbidity of the culture using the spectrophotometer.
- Spectrophotometer.
- Test tubes and flasks of the appropriate dimensions and shape for the growth of microbial cultures with aeration.

- A rotating platform or similar device at ambient temperature (between 22.0°C and 25.0°C) for agitation of the broth cultures to provide aeration.
- A hand-held illumination device that emits UV radiation with a wavelength of 254 nm.
- A room that may be made completely dark for viewing fluorescence.

Procedure

Grow the small starter culture

- Verify that all of the colonies on the MGY agar plate are PG4180 bacteria by examining them for fluorescence as explained earlier in this exercise in *Technical background*.
- Inoculate a starter culture with a colony of PG4180 bacteria. Use between 10.0 and 25.0 mL of sterile MGY medium in a test tube or flask.
- Incubate the culture, with aeration, at between 22.0°C and 25.0°C for 2.0 days.
- Verify that the bacteria that have grown in this MGY broth culture are PG4180 by assaying again for fluorescence. Aseptically remove an aliquot and examine it in the way that the colonies on the agar plate were examined.
- Quantify the cellular density of the culture as explained earlier in this exercise in *Technical background*. It should be approximately $2.0 \times 10^8 \, \text{mL}^{-1}$.

Grow the large subculture

Note: The ideal total volume of this culture is 2.0 L. The inoculated broth should be sloshing around during the incubation to ensure that atmospheric oxygen is entering the culture. It will probably be necessary to distribute the culture into more than one flask to ensure that this occurs.

- If the starter culture reached a cellular density close to the expected value, use 4.0 mL of it to inoculate 2.0 L of sterile HSC broth to create the subculture.
- Incubate the subculture with aeration, at between 22.0°C and 25.0°C for 3.0 days. This culture is *not* expected to exhibit fluorescence. To verify this, aseptically remove an aliquot and look for fluorescence as was done for the colonies on the agar plate.
- The bacteria should be growing in exponential phase during the first 2.0 days. The cellular density of the subculture may be

assayed during this incubation by aseptically removing aliquots and using the technique mentioned earlier in this exercise in *Technical background*. The *generation time* of a species of bacteria is defined in the earlier chapter in which an overview of the methods for purification of metabolites is presented. The generation time of PG4180 during the first 2.0 days of the incubation should be approximately 8.0 h. Making this measurement and doing this calculation is a way to determine if the culture may be contaminated with a different microbe.

- After 3.0 days of incubation, assay the cellular density once again. It should be approximately $4.0 \times 10^7 \, mL^{-1}$. If the density is close to this value, proceed with the next exercise.

Evaluation of results and questions to consider

- How confident are you that all of the bacteria in the starter and subcultures are the correct strain?
- What was the total number of bacteria in the subculture at the end of the incubation?

11.2 REMOVAL OF BACTERIA FROM THE CULTURE BY CENTRIFUGATION AND FILTRATION

Theoretical and technical background

For the centrifugation described here a *swinging-bucket rotor* is used. An optimal value of the *relative centrifugal force* (RCF) that is required is specified in this procedure. The definitions of swinging-bucket rotor and RCF are provided in the earlier chapter that gives an overview of the methods for purification of metabolites.

Fundamental aspects of *liquid–liquid extraction*, also known as *solvent extraction*, are discussed in the earlier chapter that is mentioned above. The organic solvent ethyl acetate is much less polar than water, therefore these two liquids are immiscible. This technique will be employed in this exercise and again in the next exercise. The pH of the aqueous bacterial supernate will be different for the two extractions because the goal is different.

In an aqueous solution that has a pH of ~9.0 the carboxylic acid group in coronatine will be deprotonated, giving this molecule a negative charge. Organic substances that do not have an acidic group will be uncharged at this pH. Extraction of such an aqueous solution with ethyl acetate removes the nonacidic organic substances.

Materials

Reagents

- The cultures of *P. syringae* pathovar glycinea 4180, a total of 2.0 L, from the previous exercise.
- A solution of 10.0 M sodium hydroxide. A volume of 4.0 mL will probably be sufficient.
- Ethyl acetate. A volume equal to that of the bacterial supernate. The pH may be assayed with indicator paper and it should be between 5.0 and 6.0.
- A solution of 12.0 M hydrochloric acid. A volume of 12.0 mL will probably be sufficient.

Supplies and equipment

- Bottles for centrifugation of the supernate. Two or four identical bottles, with a combined volume of 2.0 L.
- A centrifuge and rotor combination with the following characteristics:
  1. The rotor should have swinging buckets that will hold the bottles described above.
  2. The dimensions of the rotor and the allowed revolutions per minute should be such that a RCF of $\sim 4000.0 \times g$ at the bottom of each bucket may be generated.
- A beaker or flask with a volume of at least 2.0 L to hold the supernate between the centrifugation and filtration steps.
- Circular filters (FisherBrand), made of glass fiber, purchased from Fisher Scientific. These filters should be of the G4 grade, have a retention of 1.2 μm and a diameter of 5.5 cm. Similar filters from a different supplier may be used as an alternative.
- A Buchner funnel with an internal diameter of ~ 5.5 cm.
- A filtering flask with a sidearm that is constructed of thicker glass that is found in a typical flask. This flask will be used to create a vacuum underneath the Buchner funnel. The thick glass prevents the flask from collapsing. The volume should be at least 2.0 L.
- Plastic tubing that will not collapse when a vacuum is applied and is of the appropriate dimensions to connect the sidearm on the filtering flask to the pump.
- A vacuum pump.
- A separatory funnel made of glass, or other material that is resistant to ethyl acetate, with a volume of 500.0 mL.
- A ring stand to hold the separatory funnel.

- A beaker or flask with a volume of at least 2.0 L to hold the supernate after the extraction.
- Bottles for storage of the clarified supernate until the next exercise is performed. They should be constructed of plastic and have screw caps. If the supernate will be stored in a freezer, the volume of the bottles, and the material from which they are constructed, should be chosen to prevent cracking of the plastic.

Procedure

Removal of the bacteria from the culture

- Divide the culture into two or four equal portions and pour each portion into one of the plastic bottles that are designed for centrifugation.
- Measure the mass of each bottle. If the masses differ by more than 0.1 g, adjust the volume of supernate in each so that the masses are within this range.
- Perform centrifugation so that the RCF at the bottom of each bottle is $\sim 4000.0 \times g$ for 60.0 min at ambient temperature (22.0°C).
- If it appears that greater than 90.0% of the bacteria have been sedimented, combine the supernates by decanting them into a clean container.
- Perform vacuum-driven filtration to remove any remaining bacteria from the supernate. Use the filter, Buchner funnel, filtering flask, tubing, and pump that were described earlier in section "*Materials*." If you notice deterioration of the filter during this process, replace it with a fresh filter. The filtrate should lack turbidity.
- Assay the pH of the clarified supernate. It should close to 7.1.

Removal of nonacidic organic substances from the supernate

Note: The supernate from the bacterial culture is an aqueous solution that includes coronatine. The density of this solution is expected to be very close to that of pure water. Because the density of ethyl acetate (0.90 g mL^{-1}) is less than that of water, this organic solvent should be the upper layer in the separation step of each extraction.

- To ensure that coronatine has a negative charge that will prevent it from being extracted into the nonpolar ethyl acetate, add 4.0 mL of 10.0 M sodium hydroxide to the supernate and stir to equalize the pH throughout the solution.

- Assay the new pH of the supernate. It should be close to 9.0.
- Divide the supernate into eight batches of 250.0 mL each. Extract each batch with an equal volume of ethyl acetate. After each extraction, pour the aqueous phase into a large beaker or flask and discard the organic phase.
- The entire bacterial supernate should now be in one beaker or flask.

Acidification of the supernate

Note: In the next exercise coronatine will be extracted from the aqueous bacterial supernate into ethyl acetate. Protonation of the carboxylic acid group in coronatine minimizes the polarity of this molecule, making it more soluble in ethyl acetate than in the supernate.

- To ensure that the carboxyl group in the coronatine is pro-tonated, add 12.0 mL of 12.0 M hydrochloric acid to the supernate and stir to equalize the pH throughout the solution.
- Assay the new pH of the supernate. It should be close to 2.1.

Storage of the clarified, extracted and acidified supernate

- Pour the clarified, extracted, and acidified supernate into the bottles that are designed for storage. If the next exer-cise will not be performed until a later date, the containers should be shielded from light to prevent photolysis of coro-natine. Your instructor will decide whether to store the con-tainers at ambient temperature, in a refrigerator or in a freezer.

Evaluation of results and questions to consider

- Comment on the effectiveness of centrifugation for sedimen-tation of the bacteria.
- Comment on whether or not the filtration step was neces-sary and whether or not it decreased the turbidity of the supernate.

11.3   EXTRACTION OF CORONATINE FROM THE SUPERNATE OF THE CULTURE WITH AN ORGANIC SOLVENT

Theoretical and technical background

Fundamental aspects of *liquid—liquid extraction*, also known as *solvent extraction*, are discussed in the earlier chapter in this book that provides an overview of the methods for purification of metabolites. The organic solvent ethyl acetate is much less polar than water, and therefore these two liquids are immiscible. In an aqueous solution with a pH of ~ 2.1, the carboxylic acid group of coronatine is protonated and this form of the molecule is less polar than the deprotonated form. Although protonated coronatine remains soluble in water, it is more soluble in ethyl acetate. This metabolite may be extracted from the aqueous supernate of a bacterial culture that has a pH of ~ 2.1 into ethyl acetate.

Materials

Reagents

- The clarified bacterial supernate from the previous exercise. Nonacidic organic substances should have been removed by extraction. This supernate should have been acidified to a pH of ~ 2.1 and the volume should be ~ 2.0 L.
- Ethyl acetate: A volume equal to that of the bacterial supernate. The pH may be assayed with indicator paper and it should be between 5.0 and 6.0.

Supplies and equipment

- A separatory funnel made of glass, or other material that is resistant to ethyl acetate, with a volume of 500.0 mL.
- A ring stand to hold the separatory funnel.
- A beaker or flask, made of glass or other material that is resistant to ethyl acetate, and of sufficient size to hold the combined volume of the organic extracts (~ 2.0 L).

Procedure

Note: The supernate from the bacterial culture is an aqueous solution that includes coronatine. The density of this solution is expected to be very close to that of pure water. Because the density of ethyl acetate (0.90 g mL^{-1}) is less than that of water, this organic solvent will be the upper layer in the separation step of each extraction.

- If the clarified, extracted, and acidified supernate was frozen at the end of the previous exercise, thaw it in a bath of water maintained at 37°C.
- Divide the supernate into eight portions of ~ 250.0 mL each.
- Pour one portion of the aqueous supernate into the separatory funnel and extract it with an equal volume of ethyl acetate. Discard the aqueous phase. Recover the organic phase, which should now include the coronatine, in the beaker or flask.
- Repeat the extraction process with the remaining seven portions of the aqueous supernate. Combine each of the organic extracts with the first organic extract in the beaker or flask.
- If the next exercise will not be performed immediately, cover the container that holds the combined organic extracts and arrange shielding to prevent exposure to light that might cause photolysis of coronatine. Store the container in a fume hood at ambient temperature (22.0°C).

11.4 REMOVAL OF SOLVENT FROM THE FRACTION ENRICHED IN CORONATINE AND RESUSPENSION OF THE DRIED SUBSTANCE

Theoretical and technical background

There are two parts to the process of drying of the solution that contains coronatine. In the first part, water is removed by adsorption. In the second, the organic solvent is removed by evaporation. Fundamental aspects of these techniques are presented in the earlier chapter in this book that provides an overview of the methods for purification of metabolites.

Materials

Reagents

- The organic liquid extract, which contains coronatine, from the previous exercise. A volume of ~ 2.0 L.
- Anhydrous magnesium sulfate. A mass of 0.5 g is needed for each 1.0 mL of water that is underneath the organic liquid extract.
- Methanol for resuspension of the dry coronatine. A volume of 2.0 mL should be sufficient.

Supplies and equipment

- A beaker with dimensions that will allow for rapid evaporation of solvent. If a beaker with a volume of 4.0 L is available, this will be ideal. The diameter of the opening at the top of a beaker is the same as the diameter at the base. Do not use a container in which the opening at the top has a smaller diameter than the diameter at the base (e.g., an Erlenmeyer flask). Evaporation is very slow in this type of flask.
- Beakers with a smaller volume for the final stage of evaporation, for example, 500.0 and 50.0 mL.
- A fume hood.

Procedure

- If water is evident underneath the organic liquid, estimate the volume. Add 0.5 g of anhydrous magnesium sulfate for every 1.0 mL of water.
- After the water has been adsorbed, decant the organic liquid into the beaker that will be used for evaporation of solvent. Avoid transferring the hydrated magnesium sulfate.
- Place the beaker that contains the organic liquid extract in a fume hood and leave it uncovered to allow the solvent, ethyl acetate, to evaporate.
- When the volume of solvent has decreased to ~200.0 mL, pour the liquid into a beaker that has a volume of 500.0 mL. Allow the evaporation to continue in this smaller, uncovered beaker.
- When the volume of solvent has decreased to ~20.0 mL, pour the liquid into a beaker that has a volume of 50.0 mL. Allow the evaporation to continue in this even smaller, uncovered beaker.
- After the evaporation has proceeded for a total of ~3.0 days, no solvent should remain. The residue in the bottom of the beaker should be pale yellow.
- Resuspend the dry residue, which should include coronatine, in 2.0 mL of methanol.
- Divide the sample into several portions. The portions should be stored in tightly sealed test tubes, shielded from light that might cause photolysis of this metabolite. If a freezer is available, this is the ideal place to store these samples.

11.5 ASSAY ON THE LEAVES OF PLANTS FOR THE CHLOROSIS ACTIVITY OF CORONATINE

Theoretical background

When chlorosis occurs in the leaf of a plant the pigment chlorophyll is lost from the chloroplasts (Uppalapati et al., 2007). If chlorosis occurs in a small spot on a leaf, the loss of the chlorophyll in this spot results in a yellow spot on an otherwise green leaf. This change in color may be observed with the unaided eye. Although *P. syringae* pathovar glycinea 4180 from which the coronatine has been purified is typically associated with *Glycine max* (soybean), induction of chlorosis by purified coronatine that was applied to the leaves of *Solanum lycopersicum* (tomato) has also been reported (Palmer & Bender, 1995).

The theory that is currently being proposed is that coronatine inhibits the salicylic acid signaling that plants employ as a defense against microbes such as the various pathovars of *P. syringae* (Panchal et al., 2016; Uppalapati et al., 2007). The mechanism of this interference involves the structural similarity between coronatine and a hormone, jasmonoyl isoleucine (JA-Ile), that is expressed in plants. One of the several functions of JA-Ile is the inhibition of the salicylic acid-mediated pathway.

Technical background

In the published method for assaying purified coronatine for chlorosis-inducing activity, the toxin that has been dissolved in methanol is diluted into deionized water so that only the minimum amount that is necessary will be applied to the leaf (Palmer & Bender, 1995). The dilution also decreases the concentration of methanol in the sample, minimizing the possibility of toxicity from the solvent.

Note: Exercise 11.6 should be performed prior to Exercise 11.5. The molar concentration of the coronatine that has been purified must be determined so that the appropriate amount may be applied to the leaves.

Materials

Reagents

- For the *experimental sample*.
    An aliquot of the solution of coronatine, in methanol, that was purified in earlier exercises.
    The total amount of coronatine in this aliquot should be 7.5 nmol (2.4 μg).

- For the *positive control sample* (optional).

  You may be provided with an aliquot of coronatine that was obtained from a commercial supplier to be used as a positive control.
- For the *negative control sample.*

  Pure methanol. No more than 0.1 mL is needed.
- An *Heirloom Tomato* plant or a similar strain should be used. It should be approximately 33.0 cm tall. Leaves, but no tomatoes, should be present on the stem. If only the experimental sample and the negative control will be assayed, six leaves will be sufficient. If the positive control will also be assayed, nine leaves will be necessary. The area of each leaf will be approximately 30.0×50.0 mm by the end of the experiment. It should be in a small pot of soil.

Supplies and equipment

- Micropipettors and plastic pipette tips.
- Plastic snap-cap tubes for preparing the diluted samples.
- A metric ruler for measuring the spots of chlorosis.

Procedure

Note: For each of the two or three diluted samples to be assayed at least 15.0 μL is needed.

- Prepare the experimental sample.

  Dilute an aliquot of the coronatine that has been purified into deionized water to a concentration of 0.5 mM (0.5 nmol μL^{-1}). Calculate the final concentration of methanol in this diluted sample.
- Prepare the positive control sample (if appropriate).

  If a sample of coronatine from a commercial source is provided, dilute an aliquot as was done for the coronatine that was purified.
- Prepare the negative control sample(s).

  Dilute an aliquot of pure methanol into deionized water to a concentration equal to that of the methanol in the diluted experimental sample. If the positive control will be assayed, and the concentration of methanol in this sample after dilution is different than that in the diluted experimental sample, a second dilution of methanol at the appropriate concentration should be prepared.
- Apply aliquots of the diluted samples to the leaves.

Each aliquot applied to a leaf should have a volume of 5.0 μL. For the experimental and positive control samples, this volume will contain 2.5 nmol (0.8 μg) of coronatine.

Use a micropipettor to apply each aliquot to a distinct leaf on the plant. Apply the drop to the *underside* (not as shiny as the top side) of the leaf. After applying each aliquot, gently attach a label that shows which sample was applied. Use a twist-tie or a piece of string rather than an adhesive that might be toxic to the plant.

For each sample to be assayed, apply an aliquot to each of three leaves.

- If chlorosis is induced it may take as long as 9.0 days to become visible. The overall health of the plant should be stable for at least 13.0 days. Keep the plant in the laboratory at ambient temperature ($\sim$22.0°C). It should be exposed to light from ceiling fixtures for $\sim$12.0 h each day.
- Inspect the leaves each day for evidence of chlorosis.
- If chlorosis becomes apparent on one or more of the leaves to which a sample was applied, record your observations. Your notes should include how many of each of the three leaves that received an aliquot of a particular sample show chlorosis. Also record the diameter of each spot of chlorosis. Capture images as appropriate.

Evaluation of results and questions to consider

- For each sample that was applied to leaves, discuss whether or not it induced chlorosis.
- If the positive control sample was applied to leaves, compare these results with those for the experimental sample. Discuss whether or not the results are consistent with the conclusion that coronatine has been purified.
- Discuss whether or not the methanol in these samples might have induced chlorosis.
- Compare your results to those presented in the scientific literature (Palmer & Bender, 1995; Uppalapati et al., 2007).

11.6 SPECTROPHOTOMETRIC QUANTIFICATION OF THE YIELD OF CORONATINE USING ULTRAVIOLET RADIATION

Theoretical background

As is the case for most organic molecules that are not associated with a metallic ion, coronatine absorbs radiation with a wavelength in the near UV region but not visible radiation. The wavelength of maximum absorption (λ_{max}) has been reported as 208.0 nm and the molar extinction coefficient, ε_{208}, as 8.38 mM^{-1} cm^{-1} (Ichihara et al., 1977). In this experiment, it is possible that the λ_{max} will be found to be slightly lower or higher; however, it should be between 195.0 and 215.0 nm.

Technical background

In the exercises presented earlier in this chapter coronatine was extracted from the acidified, aqueous supernate of the bacterial culture into the organic solvent ethyl acetate. Ethyl acetate is much less polar than water. Coronatine is soluble in water up to 0.2 mg mL^{-1}. It is extracted into the organic solvent in spite of this aqueous solubility because it is more soluble in ethyl acetate. Coronatine is soluble in methanol up to 20.0 mg mL^{-1}. This high solubility allowed for resuspension of coronatine in methanol, after removal of the ethyl acetate into which it had been extracted, in an earlier exercise.

Most organic solvents absorb radiation in some part of the near UV region. To minimize the possibility that absorption of UV radiation by the methanol in the analyte will interfere with this assay, deionized water should be used as the solvent for the dilution that is described below in *Materials* and *Procedure*. Water does not absorb near UV radiation.

Materials

Reagents

- An aliquot of the preparation of coronatine that was extracted into ethyl acetate, dried and then resuspended in methanol in earlier exercises. If your technical skills in the various steps of the purification were good, a sample of 10.0 or 20.0 µL will be enough to acquire acceptable data. A greater volume will be needed if the yield was lower than expected. This aliquot of the preparation of coronatine will be diluted into deionized water to provide sufficient volume for spectrophotometry. The design of the cuvette that is used will have an effect on the volume of sample that is necessary. If the volume of solution needed to fill the cuvette to the proper level is 0.5 mL, then 10.0 µL of sample should be enough. If 1.0 mL is required to fill the cuvette, then 20.0 µL of sample will be needed.
- An aliquot of pure methanol to prepare the negative control sample. Less than 1.0 mL is needed.

Supplies and equipment

- Micropipettors and plastic pipette tips.
- Plastic snap-cap tubes for preparing the diluted samples.
- One or two rectangular cuvettes made of quartz that is transparent to near UV radiation. The volume of solution necessary to fill the cuvette to a level above the point at which the radiation will pass through should be determined before preparing the samples for the assay. For most cuvettes, 0.5 or 1.0 mL is sufficient. The distance through the solution in the cuvette that the radiation will pass through, known as the *path length*, should be measured. For most cuvettes this is 1.0 cm. If a dual-beam spectrophotometer will be used, two cuvettes will be needed.
- A spectrophotometer that is capable of quantifying absorbance of radiation with wavelengths from 190 to 400 nm, that is, the near UV. It is best if the device will scan this range and plot the spectrum by means of associated software; however, if the only device available is one that records data at a fixed wavelength, it will be sufficient.

Procedure

Set the baseline

- Prepare a cuvette with the appropriate blank solution. The coronatine is dissolved in methanol and will be diluted into deionized water; therefore, the blank should be prepared in the same manner, substituting pure methanol for the coronatine-in-methanol sample.
- Set the baseline. If the device is capable of scanning, the baseline should be set for the range 190−400 nm. If the device only does fixed wavelengths, set it at 208 nm.

Prepare the sample of coronatine and quantify the absorbance

- Dilute an aliquot of the sample of coronatine into a volume of deionized water that is sufficient for the cuvette to be used. Refer to the discussion in section "*Materials*" above regarding the appropriate volume of the coronatine to be diluted.
- Record a spectrum of the entire range of near UV, or the A_{208}, of the diluted coronatine. If the absorbance appears to be too weak to be accurate, repeat the assay with a greater aliquot of the sample of coronatine. If a greater aliquot of

the coronatine is used, then the setting of the baseline should be performed with the appropriate greater volume of methanol diluted into water.

Evaluation of results and questions to consider

- If a scan of wavelengths was performed, discuss whether or not the λ_{max} appears to be at 208 nm.
- Calculate the molar concentration of the coronatine in the preparation prior to dilution for spectrophotometry. If your technique was good, millimolar will probably be appropriate. The following should be used.
  1. The A_{208} of the diluted coronatine.
  2. The ε_{208} of coronatine that was stated earlier in this exercise.
  3. The path length in the cuvette.
  4. The Beer–Lambert law as described in the earlier chapter that provided an introduction to the use of UV radiation for spectrophotometry.
  5. The dilution ratio for the sample in the cuvette.
- Using the total volume of the preparation of the coronatine, and the concentration you have just calculated, calculate the yield in units of moles. If your technique was good, micromoles will probably be the most appropriate.
- Calculate the yield of coronatine in units of mass, using the published value of 319.18 g mol^{-1} (Ichihara et al., 1977) as the molar mass.

11.7 INFRARED SPECTROSCOPIC ELUCIDATION OF THE STRUCTURE OF CORONATINE

Theoretical background

A chapter presented earlier in this book explains how quantification of the absorption of infrared radiation by organic molecules is used to reveal their structures. The quantity that is recorded is the absorption of energy due to the stretching or bending of chemical bonds.

Technical background

Because the solvent in which coronatine is dissolved (methanol) is an organic molecule, it is imperative to distinguish absorption of energy by the bonds in the molecules of analyte from the absorption by bonds in the molecules of solvent.

Some IR spectrometers are designed in a way that a solution of analyte may be allowed to dry, eliminating the solvent, before the spectrum is captured.

Materials

Reagents

- An aliquot containing 10.0 nmol of coronatine (in methanol) will be needed for each assay with the IR spectrometer. This corresponds to 3.2 μg. If your technical skills were good during the purification of coronatine, a volume of approximately 2.0 μL should contain this amount.

Supplies and equipment

- Micropipettors and disposable pipette tips.
- An IR spectrometer that is designed to assay organic substances.

Procedure

- If the device being used is designed to assay dry samples, place a drop containing 10.0 nmol of coronatine onto the sample-loading spot. Allow the solvent to evaporate.
- Acquire the spectrum of absorption of IR radiation.

Evaluation of results and questions to consider

- The IR spectroscopic data that has been published for coronatine is a list of the wavenumbers of the bands of absorption rather than a graphical spectrum (Ichihara et al., 1977). Compare the IR spectrum that you have recorded to the published data. Discuss whether or not it is possible that coronatine has been purified.

11.8 ASSAY, BY MASS SPECTROMETRY, OF THE STRUCTURE OF CORONATINE

Theoretical background

A chapter presented earlier in this book provided an introduction to mass spectrometry (MS). As explained in that chapter, a molecule must be in an ionic form to be projected through the electric field toward the detector. The uppercase letter "M" is used to represent a molecule of neutral analyte (Niessen, 2006).

For MS, there are several ways in which a neutral organic molecule may be converted to an ion. Two of these ways are the addition to, or loss of a proton from, this molecule. This generates a *protonated molecule*, indicated by the notation + H, or a *deprotonated molecule* indicated by the notation −H. The net charge on the ionic form is denoted by a plus or minus sign in superscript. The protonated form is thus denoted $[M + H]^+$, whereas $[M−H]^-$ represents the deprotonated form. To detect $[M + H]^+$ the mass spectrum is recorded in *positive-ion mode*, whereas *negative-ion mode* is used to detect $[M−H]^-$. According to the scientific literature, coronatine may be detected as an $[M + H]^+$ ion (Ichihara et al., 1977).

Technical Background

For a pure substance, good data may be obtained by injecting it directly into an MS device. If the analyte is expected to be a mixture, or if it is possible that the purification method did not remove all of the undesired substances, the hybrid technique of liquid chromatography−mass spectrometry (LC−MS) should be used. If the chromatography parameters are set properly, LC will separate the substance of interest from other substances. In this way a mass spectrum of the pure substance of interest may be obtained.

As explained in the earlier chapter that was mentioned above, the data produced by MS is a *mass spectrum*. The symbol *m* represents the molar mass of an analyte in grams per mole and the symbol *z* represents the number of charges on each molecule of this analyte. A mass spectrum is a histogram of the relative amount of analyte molecules that have a particular *m/z* ratio.

LC−MS produces two types of data. For LC, a *chromatogram* in which the amount of analyte eluting from the column is plotted as a function of time is generated. For coronatine, a mass detector will be used to generate the chromatogram. For analytes that efficiently absorb UV or visible radiation, spectrophotometry generates the chromatogram. If there is single peak on the chromatogram, a mass spectrum of the substance in this peak is examined. If there are multiple peaks, the mass spectrum of the peak that is believed to be the substance of interest is examined.

In the procedures in exercises presented earlier in this chapter, coronatine was extracted from the aqueous supernate of the bacterial culture into the organic solvent ethyl acetate. Ethyl acetate is much less polar than water. Although coronatine is soluble in water up to $0.2\,mg\,mL^{-1}$, it is extracted because it is more soluble in ethyl acetate. Coronatine is soluble in methanol up to $20.0\,mg\,mL^{-1}$. This high solubility allowed for resuspension of coronatine in methanol, after removal of the ethyl acetate into which it had been extracted, in an earlier exercise.

You will need to dilute an aliquot of this preparation of coronatine to decrease the concentration, so that an appropriate amount will be injected into the device. Deionized water should be used as the solvent for the dilution. As explained below in *Materials*, detection of organic metabolites with MS is very sensitive. The dilution will be approximately 100-fold; thus, the concentration of methanol in the diluted solution will be approximately 1.0%. It is unlikely that this amount of the organic solvent will interfere with either the chromatography or the spectrometry.

Materials

Reagents

- An aliquot of the coronatine that was purified, dried and then resuspended in methanol in earlier exercises. Because the amount of analyte necessary to obtain good data varies depending on the type of device employed and whether the MS or LC–MS technique is performed, your instructor and the laboratory staff will have to provide advice. A typical LC–MS assay of coronatine requires only 300.0 pmol (95.8 ng) of this substance. Such an amount should be only a very small percentage of your preparation.

Supplies and Equipment

- Micropipettors and disposable pipette tips.
- Plastic snap-cap tubes for preparing the diluted samples.
- *Autosampler vials* for loading samples of analyte into the MS or LC–MS device. These vials are constructed of plastic or glass. The internal compartment may have a cylindrical or conical shape. The conical shape is preferable if there is a concern that the volume of the sample may be too small for

proper uptake into the needle that will subsequently inject the sample.
- An MS or LC−MS device connected to a computer in which an application for operation of the device and collection of data has been installed.

Procedure

Preparation of the sample of analyte

- A typical volume of analyte that is injected into an MS or LC−MS device is 5.0 μL. To prepare a sample that includes 300.0 pmol of coronatine in this volume, it will be necessary to dilute an aliquot of the preparation into deionized water. Prepare the appropriate dilution.
- Inject the appropriate volume of diluted coronatine into the device and acquire the data.

Operation of the analytical device

- If MS will be performed.

  Follow the instructions provided by the instructor, laboratory staff and the manufacturer of the device.
- If LC−MS will be performed.

  In the chapter that introduced HPLC and UHPLC that was presented earlier in this book, the characteristics of a typical chromatographic column used in this technique are described. Two protocols for the process of chromatography are presented, one for a 3-min run and the other for a 10-min run. If a column that is similar to the one described in that earlier chapter is used, the protocol for a 3-min run will be appropriate for this experiment. As explained earlier in *Technical background*, a mass detector should be used to identify coronatine as it elutes from the column.

Collection of data

- MS:

  A mass spectrum should be recorded.
- LC−MS:

  A chromatogram and at least one mass spectrum should be recorded. If multiple peaks that have eluted from the column are examined with MS, multiple mass spectra should be obtained.

Evaluation of results and questions to consider

- Compare the data that have been obtained to the published mass spectral data for coronatine (Ichihara et al., 1977). Discuss whether or not your data suggest that coronatine has been purified.
- What is the molecular formula of coronatine? Is your data consistent with this formula?
- If LC−MS was performed and multiple peaks were observed on the chromatogram, are you convinced that the mass spectrum you examined was from the peak containing coronatine? Explain your reasoning.

11.9 NUCLEAR MAGNETIC RESONANCE SPECTROSCOPIC ELUCIDATION OF THE STRUCTURE OF CORONATINE

Theoretical background

As explained in the chapter presented earlier in this book in which this technique was introduced, there are several isotopes to which NMR spectroscopy may be applied. In organic substances the most abundant of these is the predominant isotope of hydrogen, ^1H. In this exercise ^1H NMR spectroscopy of the analyte, coronatine that was purified in earlier exercises, is performed. This technique is also known as *proton NMR*.

Technical background

Refer to the chapter presented earlier in this book in which NMR spectroscopy was introduced and to some of the cited references (Richards & Hollerton, 2011) for a discussion of appropriate solvents. One of the goals in choosing a solvent is minimization of resonance peaks in the spectrum due to the molecules of solvent. For ^1H NMR spectroscopy, the analyte, coronatine in this case, should be dissolved in a solvent in which the atoms of ^1H have been replaced with ^2H (deuterium). This is known as a *deuterated solvent* and it is used because ^2H does not generate a peak of resonance on the spectrum in this type of experiment. Deuterated versions of water or methanol are often used as solvents for organic metabolites being examined with NMR spectroscopy. The disadvantage of these two solvents is that exchange between the ^1H in acidic groups in the analyte and the ^2H in the solvent results in loss of the expected peaks of resonance for these protons.

   The solvents in which the solubility of coronatine is high enough for NMR spectroscopy are deteuro-methanol (CD_3OD) and deutero-dimethyl sulfoxide (D_6-DMSO). Because of the exchange of ^1H from acidic groups in molecules of analyte with ^2H in CD_3OD that was mentioned earlier, D_6-DMSO is the preferred solvent for the siderophore to be examined in this experiment.

   Peaks of resonance due to the solvent are not completely eliminated by the use of ^2H. As explained in the earlier chapter mentioned above, some exchange of ^1H in the analyte with ^2H in the molecules of solvent will result in a solvent-specific peak or cluster of peaks in the NMR spectrum and this may complicate interpretation of the data. In an experiment in which an analyte is dissolved in D_6-DMSO, be aware that under atmospheric pressure the boiling point of this liquid is 189.0°C. After a solution in D_6-DMSO is prepared, it is difficult to increase the concentration of the solute because this solvent does not readily evaporate.

Materials

Reagents

- D_6-DMSO to dissolve the analyte, coronatine, as explained in *Technical background* earlier in this exercise. As discussed below, each assay usually requires less than 1.0 mL of solvent.
- An aliquot of the coronatine that was purified, dried and then resuspended in methanol in earlier exercises. If the frequency provided by the NMR spectrometer is 500.0 MHz or greater, 3.0 μmol (0.96 mg) of coronatine should be sufficient to obtain good data. If the coronatine is resuspended in D_6-DMSO at a concentration of 5.0 mM (1.6 mg mL^{-1}), then 0.6 mL will contain this amount of coronatine. This is a typical volume for an NMR spectroscopic experiment.

Supplies and equipment

- Micropipettors and pipette tips.
- A test tube for evaporation of the methanol solvent from the sample of analyte and resuspension in the D_6-DMSO.
- A fume hood or a centrifuge in a low-pressure chamber for evaporation of the methanol.
- An *NMR sample tube*, that is, cylindrical glass tube designed to hold a solution of analyte that will be assayed with NMR

spectroscopy. A tube with an outside diameter of 5.0 mm and a length of 18.0 cm is appropriate for a sample of 0.6 mL.
- An NMR spectrometer.

Procedure

- Consult with the instructor and/or the member of the laboratory staff that maintains the NMR spectrometer regarding the amount of analyte that will be necessary to obtain a good spectrum. Determine the number of moles necessary and the appropriate volume for the sample.
- Measure the appropriate volume of the solution of coronatine (in methanol) into a tube that will allow for efficient evaporation of solvent.
- Leave the tube uncovered in a fume hood or use a centrifuge in a low-pressure chamber to remove the solvent by means of evaporation.
- Resuspend the analyte in the D_6-DMSO and then use a micropipettor to transfer the sample into the NMR tube.
- Load the NMR tube that contains the analyte into the spectrometer and collect the data

Evaluation of results and questions to consider

- Compare your results to the published NMR spectrum of coronatine (Nara, Toshima, & Ichihara, 1997). Although deutero-chloroform was used as the solvent to obtain this published data, the NMR spectrum obtained for coronatine dissolved in D_6-DMSO is expected to be similar. Are the data from your experiment consistent with the conclusion that coronatine has been purified? Explain your reasoning.

BIBLIOGRAPHY

Ichihara, A., Shiraishi, K., Sato, H., Sakamura, S., Nishiyama, K., Sakai, R., ... Matsumoto, T. (1977). The structure of coronatine. *Journal of the American Chemical Society*, 99, 636–637. Available from https://doi.org/10.1021/ja00444a067.

Keane, P., Kerr, A., & New, P. (1970). Crown gall of stone fruit. II. Identification and nomenclature of agrobacterium isolates. *Australian Journal of Biological Sciences*, 23, 585–596. Available from http://www.publish.csiro.au/paper/BI9700585.

Nara, S., Toshima, H., & Ichihara, A. (1997). Asymmetric total syntheses of (+)-coronafacic acid and (+)-coronatine, phytotoxins isolated from

Pseudomonas syringae pathovars. *Tetrahedron*, *53*, 9509–9524. Available from   http://www.sciencedirect.com/science/article/pii/S00404020970 06145.

Niessen, W. M. A. (2006). *Liquid chromatography–mass spectrometry* (3rd ed.). Boca Raton, FL: CRC/Taylor & Francis. Available from https://lccn.loc. gov/2006013709.

Palmer, D., & Bender, C. (1993). Effects of environmental and nutritional factors on production of the polyketide phytotoxin coronatine by *Pseudomonas syringae* pv. glycinea. *Applied and Environmental Microbiology*, *59*, 1619–1626. Available from http://www.ncbi.nlm.nih.gov/entrez/query.fcgi? cmd = Retrieve&db = PubMed&dopt = Citation&list_uids = 16348941.

Palmer, D. A., & Bender, C. L. (1995). Ultrastructure of tomato leaf tissue treated with the pseudomonad phytotoxin coronatine and comparison with methyl jasmonate. *Molecular Plant-Microbe Interactions*, *8*, 683–692. Available from   http://www.apsnet.org/publications/mpmi/BackIssues/ Documents/1995Abstracts/Microbe08-683.htm.

Panchal, S., Roy, D., Chitrakar, R., Price, L., Breitbach, Z. S., Armstrong, D. W., & Melotto, M. (2016). Coronatine facilitates *Pseudomonas syringae* infection of arabidopsis leaves at night. *Frontiers in Plant Science*, *7*, 880. Available from https://www.ncbi.nlm.nih.gov/pubmed/27446113.

Richards, S. A., & Hollerton, J. C. (2011). *Essential practical NMR for organic chemistry*. Chichester, West Sussex, UK: John Wiley. Available from https:// lccn.loc.gov/2010033319.

Uppalapati, S. R., Ishiga, Y., Wangdi, T., Kunkel, B. N., Anand, A., Mysore, K. S., & Bender, C. L. (2007). The phytotoxin coronatine contributes to pathogen fitness and is required for suppression of salicylic acid accumulation in tomato inoculated with *Pseudomonas syringae* pv. tomato DC3000. *Moleculat Plant-Microbe Interactions*, *20*, 955–965. Available from https:// www.ncbi.nlm.nih.gov/pubmed/17722699.

Chapter 12

Designing your own experiments

12.1 SOURCES OF TECHNICAL INFORMATION AND GUIDES FOR DESIGNING PROTOCOLS

Sources on the Internet

Cold Spring Harbor Protocols from the Cold Spring Harbor Laboratory Press (Cold_Spring_Harbor_Laboratory, 2017).
   Current Protocols in the Wiley Online Library (Wiley, 2017).

Books

There are also several books, written for students at your level, that discuss some of the experimental techniques that will be necessary for your experiments (Boyer, 2000, 2012; Hardin, 2001; Jack, 1995; Marshak, 1996; Ninfa, Ballou, & Benore, 2010; Simpson, Adams, & Golemis, 2009). Although some of them are focused on application of these techniques to the study of polypeptides rather than secondary metabolites, these discussions of the fundamental aspects of these methods will be helpful.

12.2 CREATING METHODS TO PURIFY AND CHARACTERIZE FLUORESCENT MOLECULES THAT ARE SECRETED BY BACTERIA

Fundamentals of fluorescence and luminescence

Learn the difference between fluorescence (Jameson, 2014; Jameson, Croney, & Moens, 2003; Valeur & Berberan-Santos, 2011, 2013), chemical luminescence, and biological luminescence (Hastings & Johnson, 2003; Roda, 2011). Learn how to assay for fluorescence.

Purification and Characterization of Secondary Metabolites.
DOI: https://doi.org/10.1016/B978-0-12-813942-4.00012-7

Definitions for fluorescence

- Fluorophore.

   A molecule that emits radiation of a greater wavelength than the radiation that it absorbs. Absorption and emission for most of these molecules are spread over a range of wavelengths. As explained below, however, when the intensity of the absorption or emission is plotted as a function of wavelength, there is a sharp peak at the maximum wavelength. The range of wavelengths of electromagnetic radiation that are detected by the human eye, that is, visible radiation, is from 400.0 to between 700.0 and 800.0 nm. Radiation having a wavelength between 50.0 and 400.0 nm is *ultraviolet* (UV) and is not visible to humans. Some fluorophores absorb visible radiation whereas others absorb UV radiation. For the fluorophores relevant to this discussion, the emission is visible radiation.

   An example of a fluorophore is rhodamine. It was one of the first to be linked to antibodies and used for localizing structures in cells and tissues by the method of *immunofluorescence*. Many other fluorophores are now available for this technique.
- Wavelength of maximum excitation (ex).

   The wavelength, in nanometers, of the radiation that is absorbed by the fluorophore. For example, the maximum absorption of rhodamine is a wavelength of 550.0 nm.
- Wavelength of maximum emission (em).

   The wavelength, in nanometers, of the radiation that is emitted by the fluorophore. For example, the maximum emission of rhodamine is a wavelength of 570.0 nm.
- ex/em.

   The wavelengths of maximum excitation and emission of a fluorophore, written as a fraction.

   For rhodamine, ex/em = 550.0/570.0.
- Spectrofluorometer.

   A device that provides a source of visible or UV radiation of the wavelength necessary for excitation of the fluorophore and a detector for quantifying the energy in the radiation emitted by the fluorophore (Boyer, 2012; Ninfa et al., 2010).

Definitions for chemical and biological luminescence

- Chemical luminescence (chemiluminescence).

   The emission of visible radiation (defined above) due to the release of energy during a chemical reaction.

- Luminol.

    Has the molecular formula $C_8H_7N_3O_2$ and when oxidized emits chemiluminescence with a wavelength of 425.0 nm (violet) (Roda, 2011). It is used by forensic scientists at crime scenes to detect traces of blood that would not otherwise be observed. It is also used to detect traces of explosive chemicals.

    To detect blood, an alkaline solution containing luminol and hydrogen peroxide (H_2O_2) is sprayed on a surface. If the solution contacts droplets of blood, the iron(III) in the heme group within hemoglobin will catalyze oxidation of luminol by H_2O_2.

    To detect explosive chemicals that contain a nitro ($-NO_2$) functional group, such as nitroglycerin and 2,4,6-trinitrotoluene, the nitro group is released from the substance in question by pyrolysis or exposure to a laser. If pyrolysis is used, the nitrite (NO_2^-) ions released will oxidize luminol in the presence of sodium sulfite. If a laser is used, molecules of gaseous NO_2 will be released and when mixed with an aerosol of an alkaline solution of luminol, the luminol will be oxidized.

- Biological luminescence (bioluminescence).

    The emission of visible radiation by a living organism due to the energy released during a biochemical reaction that is catalyzed by an enzyme.

- Luciferase.

    An enzyme that catalyzes a biochemical reaction that generates bioluminescence. A luciferase comprises one or more polypeptides.

- Luciferin.

    A molecular substrate (reactant) in the biochemical reaction that generates bioluminescence. Although the luciferin may function only in this reaction, other reactants may have a role in many other types of biochemical reactions. For example, some reactions that release bioluminescence also involve reactants such as molecular oxygen (O_2) or adenosine-triphosphate.

- Luminometer.

    A device used to quantify the energy of luminescence.

Research the literature that pertains to fluorescent secondary metabolites that are secreted by bacteria

- Bacteria that establish a symbiosis with a plant are known as a *saprophyte* of the host. Bacteria that infect a plant and cause disease are known as *phytopathogens*. Each type of bacteria is most commonly referred to with a two-part name in oblique font. The first word is the name of the genus, the second is the name of the species (Johnson & Case, 2018; Tortora, Funke, & Case, 2019). The name of the genus has an initial uppercase letter, whereas the name of the species is all lowercase. It is important to remember however that not all bacteria within a particular species are genetically identical, thus several *strains* often exist within a species. In the case of phytopathogens, each strain within a species may be referred to as a *pathovar* (abbreviated "pv", not in oblique font). One of the differences that may exist between the various pathovars within a species is the species of plant that is infected. Two types of bacteria within a species that infect two different species of plants are usually considered to be distinct pathovars (Alfano & Collmer, 1996). For example, *Erwinia carotovora* pv tomato and *E. carotovora* pv tobacco.

- Try to answer the following questions regarding the literature on this topic.
  1. In which bacterial species has the secretion of fluorescent secondary metabolites been most thoroughly studied?
  2. Do most species of bacteria secrete fluorescent secondary metabolites?
  3. If one desires to discover a new type of fluorescent secondary metabolite that is secreted by bacteria, what is a good strategy for choosing the species to assay?

Identify the appropriate bacterial strain within the appropriate species

Consider the following issues.

- Will the strain be a hazard to the health of the people in the laboratory? To purify a metabolite from the supernate of a culture, at least 1.0 L of culture will be necessary. Because the goal is to study fluorescent metabolites rather than disease in humans, it will be best to avoid strains that are pathogenic to humans.
- Will the instructor be able to obtain the bacterial strain quickly enough, so that the project can be completed by the end of the term? Many strains of bacteria may be purchased from the ATCC in Virginia (American_Type_Culture_Collection, 2017). The cost may be prohibitive; however, in which case it will be necessary to obtain the desired strain from a life science supply company or a scientist at another

academic institution or research center. This issue should be discussed as early as possible to ensure that the strain is obtained in time for the work to be completed.

- How much information about fluorescent secondary metabolites secreted by the strain under consideration is already in the scientific literature and how many scientists have already studied this strain? If there are multiple publications from scientists at several distinct institutions, it may be difficult to discover anything new in a brief project. In this case it will be more interesting to identify a strain that has not yet been documented as a source of fluorescent metabolites.
- If you are considering a strain for which there are not yet any scientific reports of secreted secondary metabolites, how likely is it that this strain secretes such a substance? Do other strains within the same species secrete fluorescent molecules?

Review the published experiments regarding purification of fluorescent molecules that are secreted by bacteria and design protocols that will be practical in the laboratory in which you will be working

Consider the following issues.

- Will the materials needed for your protocols be available?
    1. The types of equipment that may be needed for the process of growing the bacterial culture and purifying the fluorescent molecule include an incubator, a centrifuge, a separatory funnel for organic extractions, plastic columns for chromatography, and micropipettors. To monitor the rate of growth of the culture, a spectrophotometer that quantifies absorbance of visible radiation is needed. To verify that a fluorophore is being secreted by the bacteria either a hand-held source of UV radiation (wavelengths between 254.0 and 365.0 nm) or a spectrofluorometer is needed as explained below.
    2. The types of supplies that may be needed include glass-fiber filters to be used in a funnel-vacuum flask assembly to eliminate bacteria from the supernate of the culture and ultrafiltration devices to be used in a centrifuge to increase the concentration of metabolite in the clarified supernate of the culture.
    3. The types of reagents that may be needed include organic solvents and materials for solid-phase extraction such as beads of silica or Sephadex coated with a substance to which the metabolite of interest will reversibly bind.

- Will any of the materials that you are considering using be a hazard to the health of the people in the laboratory?

  Phenol is a very corrosive organic solvent. Even a small drop will burn human skin immediately upon contact. Although this solvent was used for many years in organic-aqueous liquid extractions (Meyer & Abdallah, 1978), most scientists now prefer to use alternative methods that do not involve a large volume (e.g., more than 10.0 mL) of this chemical (Bultreys & Gheysen, 2000; Demange, Bateman, Dell, & Abdallah, 1988; Torres, Pérez-Ortín, Tordera, & Beltrán, 1986). While getting ideas for your protocols from the scientific literature, look for methods that do not involve extremely hazardous materials such as phenol.

- Will you have enough time in the laboratory to complete the procedures?

  Because some parts of procedures must be performed continuously, the length of each session in the laboratory is an issue. For overnight steps, will it be essential to be in the laboratory on two consecutive days or will it be acceptable to continue after a weekend break? Will the total number of hours that you plan to be in the laboratory during the term allow sufficient time to obtain meaningful results?

Plan the growth of cultures and the purification of the fluorescent molecules

- Media for growing cultures of bacteria.

  It will be best to use a *chemically defined medium* for the growth of the chosen bacterial strain. The advantages of using this type of medium, rather than a *complex medium*, are discussed in the earlier chapter in which an overview of the techniques for purification of metabolites is presented.

  Find a recipe for a chemically defined medium that is known to allow growth of the bacterial strain, or a closely related strain, that will be used. Liquid media in glass flasks are usually used unless the strain will only grow on agar medium. In addition to allowing for good growth of the strain, some other issues should be considered. The chosen medium should allow for secretion of the fluorescent metabolite from the chosen bacterial strain. Avoid using substances that are likely to give the medium color, interfere

with detection of the fluorescence of the metabolite of interest or interfere with spectrophotometric or spectroscopic
detection methods.
- Characteristics of the chosen bacterial strain(s) for verifying
the purity of cultures.

Growing the bacterial strain in an appropriate chemically
defined medium will minimize the possibility of contamination of the cultures with unwanted microbes. It is also helpful, however, to know of easily observed characteristics of
the strain of interest that may be used to verify identity.

The most obvious characteristic in this case is the fluorescence, which should be visible in a small (less than 1.0 mL)
sample of the culture. Fluorescence is most easily observed
if the metabolite of interest absorbs UV radiation. In this
case a hand-held device that emits wavelengths in this range
may be used to assay samples with the unaided eye in a
dark room. If the metabolite of interest absorbs visible but
not UV radiation, then a different source of excitation radiation will be required. In this case a *spectrofluorometer*
device may be used. A spectrofluorometer has filters that
provide a wavelength of visible radiation for excitation and a
detector that quantifies the emitted visible radiation.

If the fluorescence of the secreted secondary metabolite
of interest is known to be regulated by interaction with
another type of molecule such as a metallic ion, this may be
used to verify that the correct bacterial strain is being grown
(Cody & Gross, 1987; Torres et al., 1986). (*Note*: When
describing an organometallic compound, the organic portion
is known as the *ligand.*) Two cultures of the strain may be
grown with or without this interacting molecule and the
cultures observed for fluorescence. The strain should only
be used for purification of the metabolite if the expected
activation or repression of fluorescence by the interacting
molecule is observed.

The scientific literature may contain descriptions of easily
observed characteristics of the strain of interest, or simple
diagnostic assays that may verify the identity of the strain in
the cultures being grown (Holt, Krieg, Sneath, Staley, &
Williams, 1994; Johnson & Case, 2018; Tortora et al., 2019;
Whitman, 2015). The size, shape, and color of colonies on agar
medium are examples of such characteristics. The diagnostic
assays typically reveal species or strain-specific metabolic
processes.

Another aspect of bacterial physiology that may be used for identification is the *generation time*. This is the time required for a bacterium to replicate its genome and divide to produce two daughter cells. The generation time of a culture of a single strain of bacteria may be measured as described in the earlier chapter in which an overview of the methods for purification of metabolites is presented. If the generation time for the strain of interest has been reported in the scientific literature, the value calculated for the strain currently being used may be compared to verify identity. It is important to consider the temperature at which the culture is being grown because generation time is very sensitive to temperature.

Plan experiments to examine the structure of the purified molecule

Note: Once a substance is believed to be pure and is being subjected to structural analysis it is referred to as the *analyte*.

- Refer to previous chapters in this book regarding the various chemical techniques that may be used to examine the structure of the analyte.
- For a more thorough description of these techniques, several of the books listed in the bibliography will provide guidance. Some of these books cover the major forms of spectroscopy, that is, absorption of UV, visible and infrared radiation, and nuclear magnetic resonance (NMR) (Crews, Rodríguez, & Jaspars, 2010; Field, Sternhell, & Kalman, 2013; Taber, 2007; Vollhardt & Schore, 2014). These books also provide guidance in the use of mass spectrometry (MS). Two of the other books in this list focus specifically on NMR spectroscopy (Keeler, 2010; Richards & Hollerton, 2011). Others address high-performance liquid chromatography, that is abbreviated as HPLC (Meyer, 2010), high-performance liquid chromatography combined with mass spectrometry that is abbreviated as LC−MS (Niessen, 2006), and ultra-high performance liquid chromatography that is abbreviated as UHPLC (Xu, 2013). The references for solving molecular structures by growing crystals of purified organic compounds and evaluating single crystals with X-ray diffraction are provided in exercises in an earlier chapter in which crystals of ferrioxamine E are grown and examined.
- To decide which of these techniques to apply to the purified substance, the following issues should be considered.

1. Is the necessary equipment available? Although the most sophisticated and expensive devices that are required for the methods described above such as a mass spectrometer, NMR spectrometer, or X-ray diffractometer are not common in laboratories dedicated to teaching, the instructor may be able to have the chemical analysis done in a local research laboratory or at another institution.

2. In what solvent should the purified molecule be dissolved in order to obtain good data? For spectroscopy, the solvent may interfere with interpretation of the data. Choice of solvent may also affect techniques such as MS and HPLC. It may be necessary to evaporate the solvent in which the substance was purified and resuspend it in a more appropriate liquid.

3. What technology for detection of analyte is best to use for HPLC experiments? Absorption of UV or visible radiation is typically used to detect the analyte as it elutes from the column in an HPLC experiment. The UV−visible spectrum of organic molecules varies enough that it can aid in identification; however, simple absorption of radiation is not rare. Most organic molecules absorb at some wavelength in the UV range. Fluorescence induced by absorption of radiation or light is not nearly as common in organic molecules as is simple absorption. Therefore using a detector on an HPLC device that is capable of quantifying fluorescence, in addition to absorption of radiation, may be the best way to identify the desired analyte.

4. Is it practical to attempt crystallography? In most cases this technique will be the most challenging of the methods discussed here. It is often necessary to try several crystal-growing procedures before useful crystals are obtained and it can be difficult to predict how long this will take. The choice of solvent for the purified molecule is especially important for this technique.

5. What mass of analyte is needed? The mass required can vary quite a bit between the methods described above. A common way to quantify the concentration of an organic molecule in solution involves use of the *molar extinction coefficient*. This value is also known as the *molar absorptivity* and is symbolized with a Greek epsilon (ε). The concentration of analyte is typically in the micromolar or millimolar range when the absorbance of UV or visible

radiation is measured in a spectrophotometer. The definition of ε, however, is the theoretical absorbance at a specific wavelength (in nanometers) of a 1.0 M solution of the analyte when the radiation passes through 1.0 cm of solution. Thus the molar extinction coefficient for the phytotoxin coronatine measured at a wavelength of 208.0 nm, that is, the ε_{280}, has been reported as 8378.0 $M^{-1} cm^{-1}$. (*Note*: For convenience, the unit for the ε of metabolites in earlier chapters is $mM^{-1} cm^{-1}$.) Although the exact value for the ε of an analyte that has just been purified is often not known, it may be possible to obtain an estimate by retrieving this value for a molecule reported in the scientific literature that is presumed to be similar to that which has been purified.

BIBLIOGRAPHY

Alfano, J. R., & Collmer, A. (1996). Bacterial pathogens in plants: Life up against the wall. *Plant Cell, 8*, 1683–1698. Retrieved from <https://www.ncbi.nlm.nih.gov/pubmed/12239358>.

American_Type_Culture_Collection. (2017). *National collection of microorganisms*. Manassas, VA. Retrieved from <https://www.atcc.org>.

Boyer, R. F. (2000). *Modern experimental biochemistry* (3rd ed.). San Francisco, CA: Benjamin Cummings. Retrieved from <https://lccn.loc.gov/00044528>.

Boyer, R. F. (2012). *Biochemistry laboratory: Modern theory and techniques* (2nd ed.). Boston, MA: Prentice Hall. Retrieved from <https://lccn.loc.gov/2010036761>.

Bultreys, A., & Gheysen, I. (2000). Production and comparison of peptide siderophores from strains of distantly related pathovars of *Pseudomonas syringae* and *Pseudomonas viridiflava* LMG 2352. *Applied and Environmental Microbiology, 66*, 325–331. Retrieved from <https://www.ncbi.nlm.nih.gov/pubmed/10618243>.

Cody, Y. S., & Gross, D. C. (1987). Characterization of pyoverdin(pss), the fluorescent siderophore produced by *Pseudomonas syringae* pv. syringae. *Applied and Environmental Microbiology, 53*, 928–934. Retrieved from <https://www.ncbi.nlm.nih.gov/pubmed/16347352>.

Cold_Spring_Harbor_Laboratory. (2017). *Cold Spring Harbor protocols*. Woodbury, NY. Retrieved from <http://cshprotocols.cshlp.org/>.

Crews, P., Rodríguez, J., & Jaspars, M. (2010). *Organic structure analysis* (2nd ed.). New York, NY: Oxford University Press. Retrieved from <https://lccn.loc.gov/2009018383>.

Demange, P., Bateman, A., Dell, A., & Abdallah, M. A. (1988). Structure of azotobactin D, a siderophore of *Azotobacter vinelandii* strain D (CCM 289). *Biochemistry, 27*, 2745–2752. Available from https://doi.org/10.1021/bi00408a014.

Field, L. D., Sternhell, S., & Kalman, J. R. (2013). *Organic structures from spectra* (5th ed.). Chichester, West Sussex: Wiley. Retrieved from <https://lccn.loc.gov/2012046033>.

Hardin, C. (2001). *Cloning, gene expression, and protein purification: Experimental procedures and process rationale*. New York, NY: Oxford University Press. Retrieved from <https://lccn.loc.gov/00045312>.

Hastings, J. W., & Johnson, C. H. (2003). Bioluminescence and chemiluminescence. *Methods in Enzymology*, *360*, 75–104. Retrieved from <https://www.ncbi.nlm.nih.gov/pubmed/12622147>.

Holt, J. G., Krieg, N. R., Sneath, P. H. A., Staley, J. T., & Williams, S. T. (1994). *Bergey's manual of determinative bacteriology* (9th ed.). Baltimore, MD: Wolters Kluwer/Lippincott Williams & Wilkins. Retrieved from <https://shop.lww.com/Bergey-s-Manual-of-Determinative-Bacteriology/p/9780683006032>.

Jack, R. C. (1995). *Basic biochemical laboratory procedures and computing: With principles, review questions, worked examples, and spreadsheet solutions*. New York, NY: Oxford University Press. Retrieved from <https://lccn.loc.gov/94036165>.

Jameson, D. M. (2014). *Introduction to fluorescence*. Boca Raton, FL: CRC Press, Taylor & Francis Group. Retrieved from <https://lccn.loc.gov/2013041155>.

Jameson, D. M., Croney, J. C., & Moens, P. D. (2003). Fluorescence: Basic concepts, practical aspects, and some anecdotes. *Methods in Enzymology*, *360*, 1–43. Retrieved from <https://www.ncbi.nlm.nih.gov/pubmed/12622145>.

Johnson, T. R., & Case, C. L. (2018). *Laboratory experiments in microbiology* (12th ed.). Hoboken: Pearson. Retrieved from <https://lccn.loc.gov/2017039734>.

Keeler, J. (2010). *Understanding NMR spectroscopy* (2nd ed.). Chichester, UK: John Wiley and Sons. Retrieved from <https://lccn.loc.gov/2009054393>.

Marshak, D. R. (1996). *Strategies for protein purification and characterization: A laboratory course manual*. Plainview, NY: Cold Spring Harbor Laboratory Press. Retrieved from <https://lccn.loc.gov/95069760>.

Meyer, V. (2010). *Practical high-performance liquid chromatography* (5th ed.). Chichester, UK: Wiley. Retrieved from <https://lccn.loc.gov/2009052143>.

Meyer, J. M., & Abdallah, M. A. (1978). The fluorescent pigment of *Pseudomonas fluorescens*: Biosynthesis, purification and physicochemical properties. *Microbiology*, *107*, 319–328. Retrieved from <http://mic.microbiologyresearch.org/content/journal/micro/10.1099/00221287-107-2-319>.

Niessen, W. M. A. (2006). *Liquid chromatography–mass spectrometry* (3rd ed.). Boca Raton, FL: CRC/Taylor & Francis. Retrieved from <https://lccn.loc.gov/2006013709>.

Ninfa, A. J., Ballou, D. P., & Benore, M. (2010). *Fundamental laboratory approaches for biochemistry and biotechnology* (2nd ed.). Hoboken, NJ: John Wiley & Sons. Retrieved from <http://www.wiley.com/WileyCDA/WileyTitle/productCd-EHEP000092.html>.

Richards, S. A., & Hollerton, J. C. (2011). *Essential practical NMR for organic chemistry*. Chichester, West Sussex, UK: John Wiley. Retrieved from <https://lccn.loc.gov/2010033319>.

Roda, A. (2011). *Chemiluminescence and bioluminescence: Past, present and future*. Cambridge, UK: Royal Society of Chemistry. Retrieved from <https://lccn.loc.gov/2012397475>.

Simpson, R. J., Adams, P. D., & Golemis, E. (2009). *Basic methods in protein purification and analysis: A laboratory manual*. Cold Spring Harbor, NY: Cold Spring Harbor Laboratory Press. Retrieved from <https://lccn.loc.gov/2008031690>.

Taber, D. F. (2007). *Organic spectroscopic structure determination: A problem-based learning approach*. New York, NY: Oxford University Press. Retrieved from <https://lccn.loc.gov/2006035525>.

Torres, L., Pérez-Ortín, J. E., Tordera, V., & Beltrán, J. P. (1986). Isolation and characterization of an Fe(III)-chelating compound produced by *Pseudomonas syringae*. *Applied and Environmental Microbiology, 52,* 157–160. Retrieved from <https://www.ncbi.nlm.nih.gov/pubmed/16347102>.

Tortora, G. J., Funke, B. R., & Case, C. L. (2019). Microbiology: An introduction, (13th ed.). Boston, MA: Pearson. Retrieved from <https://lccn.loc.gov/2017044147>.

Valeur, B., & Berberan-Santos, M. N. (2011). A brief history of fluorescence and phosphorescence before the emergence of quantum theory. *Journal of Chemical Education, 88,* 731–738. Available from https://doi.org/10.1021/ed100182h.

Valeur, B., & Berberan-Santos, M. N. (2013). *Molecular fluorescence: Principles and applications* (2nd ed.). Weinheim, Germany: Wiley-VCH Verlag GmbH & Co. KGaA. Retrieved from <https://lccn.loc.gov/2014395548>.

Vollhardt, K. P. C., & Schore, N. E. (2014). *Organic chemistry: Structure and function* (7th ed.). New York, NY: W.H. Freeman and Company. Retrieved from <https://lccn.loc.gov/2013948560>.

Whitman, W. B. (Ed.), (2015). *Bergey's manual of systematics of archaea and bacteria*. Wiley Online Library: John Wiley & Sons, Inc. in association with Bergey's Manual Trust. Retrieved from <http://onlinelibrary.wiley.com/book/10.1002/9781118960608>.

Wiley. (2017). *Current protocols*. Retrieved from <http://www.currentproto-cols.com/WileyCDA/>.

Xu, Q. A. (2013). *Ultra-high performance liquid chromatography and its applications*. Hoboken, NJ: John Wiley & Sons Inc. Retrieved from <https://lccn.loc.gov/2012035740>.

Index

Fraction enrichment, solvent removal from, 92−93
Free induction decay (FID), 71
acquisition as ^1H nuclei, 71
Frequency
domain plot, 71
of resonance, 70−72, 75
Funnel-vacuum flask assembly, 177

G

Generation time, 23, 179−180
Glucose minimal-salts and amino acids broth (GSAA broth), 108, 110*t*
Glucose minimal-salts broth, 108
Glycine max. See Soybean (*Glycine max*)
Glycolysis, 9
Goniometer, 81
Ground spin state, 135
Growing crystals
of ferrioxamine E, 138−141
process, 139
GSAA broth. See Glucose minimal-salts and amino acids broth (GSAA broth)
Gyromagnetic ratio, 69−70

H

^1H nuclear magnetic resonance spectrum (^1H NMR spectrum), 72
spin−spin coupling effects on, 73
^1H nuclei
acquisition of free induction decay as, 71
chemical shift calculation for, 72−73
external magnetic field on
continuous application effect of, 68−70
pulsed application effect of, 70−71
Hands, protection for, 3
Hazardous chemicals, minimizing exposure to, 5−6
Hazards from equipment, 3−5, 4*t*
Heirloom Tomato plant, 159
High-performance liquid chromatography (HPLC), 49, 124−125, 180
chromatographic parameters, 54−55

sample preparation, 53−54
theory and practice of, 49−53
detection and quantification of organic metabolites, 52−53
history and variations, 49−50
HPLC application to organic metabolites, 50−51
stationary phase, mobile phase and chromatographic parameters, 51−52
High-pressure liquid chromatography. *See* High-performance liquid chromatography (HPLC)
Hoitink-Sinden optimized for production of coronatine (HSC), 109, 147−148
medium, 149*t*
HPLC. *See* High-performance liquid chromatography (HPLC)
HSC. *See* Hoitink-Sinden optimized for production of coronatine (HSC)
Hydrogen peroxide (H_2O_2), 175
Hydroxamate, 11−12
Hydroxamate-containing siderophores, 128−129
Hydroxamic acid, 11−12
groups, 107, 124−125

I

Illness, 6
ILPI. *See* Interactive Learning Paradigms Incorporated (ILPI)
Immiscible liquids, 25
Immunofluorescence, 174
Incubation of bacterial cultures, 19−21
Infrared (IR)
band, 42
chromophore, 41
radiation, 41
absorption by organic molecules, 41−42
format of spectrum of absorption, 42
spectroscopic elucidation
of coronatine structure, 163−164
of siderophore structure, 133−134
of VAI-1 structure, 99−100
spectroscopy, 41
Injury, 6

Solid-phase extraction (SPE), 26,
113–114, 177
of metabolite from supernate of
culture, 26–27
Solvent extraction. *See* Liquid–liquid
extraction
Solvent removal
by evaporation, 117
from fraction enriched in coronatine
and resuspension, 156–157
from fraction enrichment, 92–93
and resuspension of dried
substance, 115–119
Soybean (*Glycine max*), 158
Space group identification of unit cell
in crystal, 82
SPE. *See* Solid-phase extraction (SPE)
Spectrofluorometer, 174, 177
device, 179–180
Spectrophotometer, 21, 35, 98, 162
Spectrophotometric assay for
chelation, 122–124, 124*t*
Spectrophotometric quantification
of VAI-1, 96–99
of yield of coronatine using UV
radiation, 160–163
of yield of siderophore, 119–121
Spectrophotometry, 35, 94, 108–109,
120, 123
evaluation of spectra as, 39–41
Spectroscopy, 35
evaluation of spectra as, 38–39
Spill of chemical, 7
Spin, 68
Spin angular momentum, 68
Spin quantum number (*I*), 68
Spindle, 24
Spin–spin coupling effects on ^{1}H NMR
spectrum, 73
Starter cultures (broth), 20–21, 147
for fluorescence, 148
growth, 150
Stationary phase for HPLC, 51–52
Streptomyces antibioticus, 11–12,
138–141
Streptomyces pilosus, 11–12, 74–75
Styrene-divinylbenzene resin,
113–114
Subculture, 20–21, 147
growth, 150–151
Supernate of culture
acidification, 154
by adsorption onto solid phase,
113–115
from bacterial culture, 91–92
coronatine extraction, 155–156
liquid–liquid extraction of
metabolite from, 25–26
nonacidic organic substances
removal from, 153–154
SPE of metabolite from, 26–27
storage of clarified, extracted and
acidified, 154
Swinging-bucket rotor, 24, 88, 111,
151
Symbiosis, 10

T
Tagetitoxin, 62–64
Technical information and guides for
designing protocols, 173
Teratogen, 5
Tetramethylsilane (TMS), 72
TFA. *See* Trifluoro-acetic acid (TFA)
Thionyl, 43–47
Time domain plot, 71
Time of retention, 51
Time-of-flight mass analyzers (TOF
mass analyzers), 59, 61
Titanium (Ti), 33–34
TMS. *See* Tetramethylsilane (TMS)
TOF mass analyzers. *See* Time-of-
flight mass analyzers (TOF mass
analyzers)
Tomato (*Solanum lycopersicum*), 14,
158
Toxic chemicals, 6
terminology pertaining to, 5
Transitory field, 70
Transmittance, 21, 34, 42
Trifluoro-acetic acid (TFA), 51–52
2,4,6-Trinitrotoluene, 175
Turbidity quantification, growth
monitoring of bacterial cultures by,
21–23
cellular density of culture
from spectrophotometric data,
21–22
dilution of aliquot of culture and
correction, 22–23
minimal generation time of species
of bacteria, 23
Two-pan balance, 4

Printed in the United States
By Bookmasters